GW01466090

**A Brain Tumour's Travel Tale**

www.auntymbraintumours.co.uk

To mum, dad, Camilla and Emma

For being my rock

**Content**

## Acknowledgements

I thank God, who has brought me through the good the bad and the ugly. I wanted to give up on him and on myself on many occasions but he told us 'Cast all your anxieties on him, because he cares for you'

Thank you to my friends and family for your patience and your love. Thank you so much to Doctor Minhas and his surgical team at St George's Hospital. Thank you to the Croydon Community Neurorehabilitation Team, Doctor von Oertzen, my seizure specialist and Attend ABI (Acquired Brain Injury) support team.

Thank you to everyone on Aunty M Brain Tumours for allowing me to do something I am very passionate about.

## Introduction

I'm not going to lie to you, I am no Jane Austen. However, I'm hoping you will enjoy finding out a bit about me and finding out about my Intraventricular meningioma (that is a tongue twister if I ever heard one). I was born on Wednesday 27th March 1983 at around 12:00 noon.

I was two months premature and only 4lbs 5oz. Aahhh! Tiny – right? Anyone who knows me would know I certainly made up for my shortcomings as I am 5'11" now. The rushing into everything didn't really change much in my life. Well, until Wednesday 21st May 2008 when I was diagnosed with a brain tumour.

I wanted to introduce a few people as you will see then in the book. Starting with my parents and my brother. Mum and Dad met in a bikers' club and are both huge bike enthusiasts. They were married in 1973 and both had good careers. My mum was in Media and Events and my dad was a Money Broker up in the city. They had my brother in 1981 and moved to what is now our family home in London in 1983. Ian my big brother (aka 'Smelly'; he calls me the same thing so it is not as bad as it sounds and also we really don't smell, well I hope not anyway), Ian moved to Sydney in 2007 after he married a Kiwi named   Unfortunately, they are divorced now but he still lives in Sydney and has a beautiful girlfriend who I hope to call my sister-in-law one day..

We have lived next door to the same families all the time I was growing up. Our road is very much like Ramsey Street in the TV series 'Neighbours' and we are very lucky to have such a close knit community.  There are four houses with families that have always lived there.  Our neighbours on the right side had two children, Emma and Blake, who were the same age as Ian and me. Ian and I have grown up to be especially close to them.  Emma is more like a big sister as she is a couple of years older than me.

Ian and I both went to boarding school when we were young. I started at Durlston Court School when I was eight from 1991 until 1996. Whilst at Durlston I met Camilla (aka 'Milly-Moo').  The funny thing is that we didn't always hang out together in Durlston; we hung around with different people.  It was after Durlston that we became very close friends – mostly on the phone as my new school was in Somerset and Camilla lived in Reading.  We stayed in contact by letter and telephone even though our next schools were miles apart.  We used to talk for an hour every night.  Most of the time I spent so long talking with Camilla that by the time I needed to call my parents, I was all talked out hee hee! Camilla still lives in Reading and is very successful as an HR Manager. After being friends for over 20 years we have been through a lot together but we will definitely get old together and sit on a  bench moaning about the weather lol.


From 1996-1999 I attended another boarding school called Millfield which is in Somerset. I went because I wanted to be at the same school as my brother and Millfield is a great sports school where I could advance my sports and get involved in competitions. After my GCSEs I decided to stop the sports side of things and focus on doing Photography and Art. I chose to be nearer to home for my A-Levels and went to Coulsdon College. I was there for two years until I was 18. I then went on to The American InterContinental University in London. I went to do a Visual Communication Degree and a Photography Degree.


**HANG ON** – I have totally strayed off the point now, sorry, where was I? Oh yes, I was going to tell you about Roland. We met while I was in college. We dated for many years, but we realised we were better as friends. Roland was really there for me when I got sick. For all his faults, one being that he owes me thousands of pounds, he has always been a really close friend who would always help if you got sick or were just going through a hard time. He played a huge part when I was diagnosed.

Becca or, as I called her, 'BecBec', is somebody I met while I was in college. I met her through her sister, Naomi Z, who was my college friend and is another close friend of mine to this day. Becca and I were inseparable for around seven years and used to do everything together. If we weren't going out on the tiles for the night, we would be babysitting for some spare pocket money. We even worked together at one point in O'Brien's Sandwich Bar while at college. We formed a great friendship and were all but joined at the hip.

I have many friends from my time at Petro-Canada, an oil and gas exploration corporation. I worked there for two years from December 2007 until December 2009. Unfortunately the London office closed down when it merged with a company called Suncor. Everyone was made redundant or transferred to other locations in Aberdeen and around the world. The company was so supportive when I got sick in May 2008 and again when I came back to work after my surgery. I have made some lifelong friends at Petro-Canada. One-being Amanda, who left work for maternity leave around the same time that I had to leave for my operation. When we both came back to work, we were able to lean on each other for support as it was strange for both of us getting back to a normal working day. We have stayed great friends and Amanda is another person who has been very supportive during my recovery

A few other people you will see in the book include Reka. We have been friends since collage. Reka is from Hungary but resides in London. We spend most of our time laughing at each other as I can't understand her accent and she can't understand me lol. Reka married Tony last year and she has been a great friend.

Finally, I will tell you about my company Aunty M Brain Tumours. In May 2011 I was not very well and was stuck at home for a long time, I had realised through social networking (Facebook, Twitter, blogs) that there was no real support for post-op sufferers like me that was fun and not too intense. I am fully aware that brain tumours are not fun but we all need to find some positives in such a devastating illness. I found a way to support people and give them information. People call me 'Aunty M' now and it is nice. I have met so many amazing people and have made wonderful friends who have been affected by a brain tumour in some way. I hope Aunty M Brain Tumours will slowly become an international support network. (Yes, I know I have big plans lol)

OK, so! Those are some of the people you will see in my diary. I have always had diaries since I was 6 and I tried to keep up with it even when I was unwell. So I am going to share my journey through my diary entries with you. I won't write everything as there is 5yrs worth, it will get boring for you so I have just pulled out the relevant bits and turned it into a story.

Just to set the scene... I did some travelling in Australia in 2006 and when I came back I got a job with Origin HR in April they are an in-house recruitment company. I was put into the Dresdner Kleinwort account, I was enjoying my life and there were no complaints.

## 1. Story of a Headache

It was June 2006, and I had just come back from being in Australia for three months. I was looking for work in London and Origin HR, an HR outsourcing and resource management service, approached me. I was brought in to work on the Dresdner Kleinwort account, in the recruitment team. I was assisting six recruiters who all dealt with different jobs within the bank.

The first year was great, and I was very happy there. In the second year, I was going on holiday to see my brother in Sydney in August for two weeks. I had a new boyfriend called William. I was very content with my life, and then...one morning in April 2007, I was on the train during the rush hour going to work. One minute I was strap-hanging with the best of them, and the next I had fainted. Two men helped me off the train and onto the platform. One of the men said, "You may have an iron problem." I thought "Yeah, that must be it." The other man said, "You might be PREGNANT!!" I laughed and said "Not unless it was by immaculate conception."

All of the passengers and the train driver were peering out the windows at me and wondering what was going on, impatient to get on with the day. I was wondering what was happening too.

Embarrassed doesn't even begin to cover it, mortified is how I felt. I just remember willing the train and all those gawping faces to move off!

I called my mum and asked her to pick me up and take me to the doctor. While I waited for her, I called work to explain what had happened and say that I would be late. I called my boyfriend William, even though we had only been dating for a month or two. I hoped he wouldn't be too put off, but he was always really understanding about it. I thought I was losing my marbles. I told the doctor and he said it must be an iron problem, and gave me some ferrous sulphate tablets.

Then, one weekend in August 2006, I was driving down to see Will and to meet his dad for the first time. His parents live and work in Dubai and his dad was over in the UK for a visit. I drove down the motorway in heavy traffic, eventually arriving in the countryside. There was a huge festival/concert going on near his house. It took me around four hours to get there instead of the hour I'd expected, but I was quite happy sitting in the traffic, listening to music. It was a bit of an experience and very entertaining watching people, some of whom had given up waiting in their cars and had started to walk to the concert – all in high spirits, some quite literally!

My eyes went a bit funny while I was sitting in the car, and I thought that I just needed to stand up for a bit. I noticed that my hands were shaking and put it down to being hungry. Eventually, I made it to Will's house. I felt really bad about taking so long to get there. I was trying to seem relaxed as Will did all the introductions with his dad. We went to a pub, but I couldn't hear anything people were saying. At times like this, you really want to make a good impression on the parents and not look too crazy. I kept smiling and hoped for the best!

After that weekend, I took a two-week holiday to visit Ian and Melissa in Sydney. I was so looking forward to seeing them. By this time, I still had the shakes and was quite bad tempered, but I was not getting headaches. Ian and I had a great time over four days at the Gold Coast, Surfer's Paradise. It was the first time we had spent that much time together since we were children and it should have been good just to hang out. I could feel myself getting stressed about really silly things, but I tried to make the most of the break and didn't tell anyone how I was feeling.

When we returned to Sydney, it was a public holiday and Ian, Melissa, a friend and I packed our bags and jumped into the car for a camping trip in the bush. It was really good fun. My eyes would occasionally go fuzzy, but I thought it was probably just my contact lenses not coping with the hot weather. After 14 days, I said goodbye to Ian and Melissa and came back to London.

In September 2007, I was sitting in the office. We were all going out for a big dinner and to an awards ceremony. I said to my supervisor that I wasn't feeling too good and that my vision was a bit strange. I thought I should go home. She said it was really important that I be there and that I would be fine (I think I could have done with assertiveness lessons at this point). I just couldn't see properly. It was just like having tunnel vision, my head in a box. All I could see was through a small area of tunnel vision to the front, with no peripheral vision. I asked God for help to get me through the night. I found that if I sat still my vision would start to come back.

My boss wanted to chat. She asked, was I happy at work? I said yes, but my headaches were getting worse. I was always having to snack when I felt faint and taking migraine tablets frequently. My boss said I should go back to the doctor, as there was obviously a problem.

I went to the doctor first thing the next morning who suggested I might be stressed and advised me to take a week off work. I hated it. I really don't like letting people down. I explained to my boss and

I was feeling very bad about the whole situation. I stayed at Will's house for five days, as it was quiet there during the week. I spent a lot of time watching TV and random movies. I was so bored and aware that I must seem to be a bit crazy, but I couldn't explain anything. I was beginning to get depressed. I felt I had let my work down. I even left my Church because I just felt like such a disappointment. I knew they would never think that, but I was so angry at myself and so unhappy. I knew that God would never leave me. But, I had left myself, I had lost my confidence.

Just before Halloween, Will and I decided to go to a Fright Night in Thorpe Park. We got there quite early, considering the park would be open until late. We went on one ride and I was left literally shaking. I didn't know why, because I loved rides. I didn't want to look bad as it was my idea to go there, so we went on another and I could feel that my body did not want to be there. My migraine came on and I thought it was just from stress and the loud music in the park. Knowing the park would be open for some time, we went back to the car and had a short sleep, hoping that things would feel better afterwards. That seemed to help, so we went back into the park and decided to go into the haunted house. All we had to do was walk through the house and past actors who jumped out at us. Easy. I am such a scaredy cat normally, but this night all I was focused on was the headache and the pain and thinking "If you touch me I might just bite back." This time, the long one-hour queues were great for me because I knew I was not looking forward to more rides. It was quite scary not to have any control over my nerves. I knew that I was being a pain in the backside. I really couldn't understand what was wrong with me.

I went to work the next day and I was still moody, emotional and stressed all the time. In November I spoke to my boss and said I thought I was not right for the job as I was really unhappy with all my problems. I left at the end of the month. I had already booked a holiday with Will to Dubai to visit his parents. I thought that it would be a great opportunity to get away from work and have a new start when I came back. I went to the doctor first and he agreed that it was probably stress and that the holiday would be a great idea, but that I should relax and shouldn't over do it. It was so frustrating because at the surgery, I never saw the same doctor. I was always seeing a different one. It felt like I was just a number and they were not really listening to what I was saying.

In Dubai, I loved the sun. The late mornings helped as well, as my headaches seemed to happen mostly early in the morning. I found it hard to relax though and spent time just lying in the sun always wondering what was going on in my head. I was really unhappy with myself.

One day we woke up early to go with Will's mother to visit her company. His mum is the head of nursing in Dubai at the Canadian Hospital. While we were waiting to meet her, we went outside for a bit, and I was not feeling right at all. As we were walking back into the hospital, I felt very sick and faint. I was trying so hard to relax but, I knew what was about to happen – AGAIN!! The next thing I knew, I was on the floor with Will trying to pick me up with one hand and holding all our bags in the other.

It was just like the first time I had fainted on the train. Both times I had a really quick dream – nice ones doing something like shopping, and then I opened my eyes and...I was just embarrassed. I was so thankful I was at the hospital already and, with Will's mother being the head of nursing, I was really well looked after. They took lots of blood tests. I truly think God made the doctors realise I needed an X-ray of my head. I told them I had bad migraines. I was asked the usual, could I be pregnant or was I anaemic. I was becoming so unhappy and fed up with feeling like it was my fault and all in my head or that I had been stupid enough to get pregnant or not look after myself and allowing myself to become anaemic. I knew I was very thin then but not from dieting. I was so not enjoying myself and was thoroughly fed up and I'm sure Will was as well at this point. I just wanted to crawl under a rock and hide and sleep, then wake up back in my home.

I had spent some of the holiday emailing employment agencies regarding new opportunities and I set up three interviews, ready for when I came back. From the three interviews, I was really interested in two of the companies. After three interviews with Brown-Forman and two with Petro-Canada I was very excited when they both offered me jobs. I decided Petro-Canada was a better choice. I started on December 17th 2007 and was so happy with my decision. I was providing administrative and secretarial support to a floor of 30 people including five MDs. Petro-Canada was a crown corporation of Canada in the field of oil and natural gas. It was headquartered in the Petro-Canada Centre in Calgary, Alberta. I was working in the London office at London Bridge. I had the most amazing view over the River Thames. I felt really blessed for the opportunities I was getting.

The job was going great, but in my personal life, I realised my memories and my recollections of everything were starting to disappear and I was feeling very helpless and depressed. I couldn't separate my thoughts and was constantly angry inside. It was terrifying. I was scared.

At the beginning of December, I went to Will's friend's wedding. The people were very hospitable and the wedding was as lovely and romantic as a wedding should be. After the ceremony I said I needed to go to bed as my head was thumping as always. I couldn't move. I stayed in bed all night and then couldn't wait until the morning and go home. I could see Will was irritated by me and I felt I had let him down for not making more of an effort with his friends. I dropped him off and then drove back home to London. I was starting to feel spaced out again and was getting beyond fed up with myself and everyone else. I went to a family friend's wedding on 30th December and took Will with me, and again I needed to sleep all the time because I was always tired and getting migraines. The relationship was not working anymore. I was tired of feeling guilty for getting ill.

Around Christmas, I was at my worst. I started waking up in the middle of the night and having the most agonising headaches. I had never felt anything like it. It was like somebody was grabbing my head and slamming it against a concrete wall. I tried the usual tablets, but nothing touched the pain. I could only cry and pray that it would stop soon. I was in so much pain, I would be sick. Night and day were just as bad as each other. I would dread waking up and just hoped that if I could get up early enough, the pain would go before I went to work. I have to say, with everything that was going on with my body I wanted to stay away from everyone. It was clear over Christmas that Will and I were not getting on anymore. We broke up at the beginning of January.

All of my efforts were poured into my new job. I really liked the people I worked with, and I would always try to go for a quick drink after work, never staying too late as I knew it could easily lead to a headache, and I didn't want anything to go wrong. I managed to handle all the strange things my body was doing and tried to adapt.

In January I went out with Naomi and Becky to a roller disco. I was really looking forward to it but when I got on the roller skates I had never been so unsteady. I was really, really uncomfortable like I would fall down all the time. My stomach was spinning. I had to hold on to something. My body was quivering. It was so weird.

I had to come up with an excuse to leave as I didn't want to explain how I felt. Luckily Naomi's sister was there and they wanted to leave earlier and go to a bar nearer home. I went with then and was fine once I was off the skates.

My shaking hands got worse during March, and, as the doctors couldn't find anything wrong other than my migraines, they thought it must be stress again. I decided to go on holiday for 11 days with Emma to Dubai, because we knew it would be very hot and we had friends out there. We also thought it would be a great break as both our relationships had recently broken up and it was coming up to a bank holiday and my 25th birthday. We had a great time, but Emma and I soon realised I was becoming very aggressive. It was never a problem, but it was noticed by both of us. I was getting slightly worried about my behaviour. I felt like I was boiling up really easily and not coping with noise or conversation.

One evening we went out and met some friends. We had planned to spend the day with them, but I'd needed to sleep as I felt drained and very weak. I thought I would be OK for that evening and said that we would find them later in the town. When I woke up, things felt a lot better. I made sure that I always had water with me and assumed that the heat made me weak, or maybe I really was anaemic. We had another great evening and Emma and I had a great holiday despite me saying some strange things. Back home in the UK, my aggressive behaviour escalated. I was getting angry so easily. I found it simpler just to stay away from most people so that I didn't offend anyone.

At the end of April my parents went to Australia to see my brother in Sydney and my uncle in Adelaide for five weeks. I said goodbye in the morning and then went to work. I didn't really want to tell them of all the strange things that were going on as I didn't want them to worry when they were on holiday. I decided to go back to Church as I was just looking to God for help. I spoke to my Pastor's wife and explained what had been going on. Sister Gill was concerned and we prayed over the situation. I thought I was having a breakdown, because even though I had no idea what that feels like, I thought it must be similar. I just couldn't explain what was happening. My memory was getting worse and I couldn't even remember if I had eaten meals. I was finding it hard to hold a conversation because I would forget what we were talking about. It looked like I was just tired but I knew it was something else.

One Saturday, I went out dancing with Becca. I was really spaced out again. My vision was strange, although it did eventually go back to normal. We danced the evening away, and then went home. The next morning, I didn't go to Church because I felt I had a bad attitude. I was fed up with my headaches every day. Becca and I had planned to drive to Brighton but I could feel a headache coming on. I gave Becca my car keys and said she should drive. I had travelled that road for years, and directed her to what I thought was the easiest route from home.

An hour later, we were lost. We did eventually make it to Brighton and had lunch on the beach. I couldn't believe I had got the directions so wrong. Luckily we made it home. I had planned to have an early night, but I was just seething with frustration and found myself phoning Camilla and telling her that I was really angry at her. I don't know what came over me, I just snapped and I had never done that before. My hands were shaking. Camilla was desperately trying to find out what was wrong, but all I could do was gibber on. By the end of the call, I managed to pull myself together. I said sorry and that I didn't mean to snap at her. It was strange, but it made me remember another conversation with Camilla about a month before when we had been chatting and I had suddenly started talking gibberish. I put it down to being tired, as usual. Things were becoming really scary. I didn't know who I was.

When I was at work I would use my hands a lot to express myself, which I have always done and used now to try to cover up words that wouldn't come out right. One day it was terrible. I remember working on a Friday and wanting to go home, but I was determined to make it through the day. All of my bosses were going to a big meeting abroad, and would be away the whole of the next week. I was talking with a colleague and having a giggle but I was in a lot of pain from a migraine. I still can't believe I made it through the day and was able to get all my work done. I had arranged for my friend Reka to come over for dinner, but realised that I could barely move, let alone cook. Reka was waiting at the door for our girly night in. Reka was happy to cook dinner and I had already put the food in the fridge in the morning. I lay on the sofa and just waited for the pain to go away. Then I was feeling a bit better, and we watched a film called 'PS I Love You'. This was a random selection, and it was the first sad and depressing film I had watched in months. I had been avoiding them after I broke up with Will.

The film is about a young lady, Holly, and her husband, Gerry, who were a happily married couple until he succumbed to a brain tumour. Even though Holly was broken hearted, she was helped through it all by her friends. It really made me think about brain tumours and hospitals and friendships. I loved it, though I really didn't know why because it was depressing. I watched it three times with different friends. I found the film really calming – somehow I think I knew something was going to happen but that I didn't need to worry about it.

On Monday 19th May 2008, I woke up at the normal time and, knowing that there were no managers around at work that day, I was able to make sure I had everything sorted on my desk. I told my colleague Abu that I was going to the doctor on the Tuesday morning. I was determined to get to the bottom of the headaches, the shaking hands, moods, my vision – just everything. I said I should be back to work in the afternoon. I was praying about everything every second of the day.

The following day I woke up early, waiting for the doctor's surgery to open at 08:30. I called my parents and Ian, as it was Ian's 27th birthday. I thought it was pointless to worry them yet; they were thousands and thousands of miles away. I said I was going to the doctor, but didn't go into details. I just said that I had a migraine, wished Ian a Happy Birthday and put the phone down. I managed to get an appointment for 09:30.

I drove down to the clinic and went in for my appointment. I could feel that I was being rude and demanding, but I was not in the mood to be shrugged off like every other time.

The doctor looked at the list of symptoms I gave her and she said, "I don't know what is wrong with your vision, but let's get your periods sorted first and have your contraceptive implant taken out of your arm as that may explain your moods changing." The implant had been inserted in October 2007 to try and regulate my periods, but it wasn't doing me any good. I said this was not enough because I had the headaches before my implant (the implant has rods that release a steady dose of one hormone only – progestogen). The doctor insisted I should have the implant removed first and we could sort the other things after. I didn't want to wait and decided to go to Croydon to get my eyes checked by my usual optician.

There was time to spare, so I took my car to get a wing mirror that had been hit the week before sorted out. On the way there my vision went funny – it kept blacking out and then coming back all in the space of a millisecond. I had noticed that had happened once or twice when I was at work a few days before, but thought it was because of my weekly sunbed I had been having to top up my tan from Dubai. I blamed myself and decided that I would never go on a sunbed again. I was really calm even though I had no idea what was going on, and I just prayed I didn't hit anything.

My vision came back and I called a friend, Roland, who had turned up in England the day before from Hungary and was at home for two weeks. I think that when you are still and alone, you suddenly realise that God has already given you all the tools that you need for life. Just open your eyes. I felt capable of handling anything and had everything I needed. Roland was only two shops away from me for his own eye check-up. We met for a quick coffee, as our appointments were both in 30 minutes. I told Roland what had happened and he said to call him after my appointment to let him know how it went.

The optician and a lady took some X-rays of my eyes and checked them. The optician asked for a second opinion. I could see she looked concerned. She said that she thought that it might be cysts in both eyes as there was bleeding in the back of them. She asked me to go straight to St George's Hospital and said that she was going to call them to expect me. I called Roland and I told him what was going on and he came straight over and took me to St George's. Although he could see that my situation was not looking good, he spent most of the time trying to make me laugh. I phoned my work colleague Abubaker again and explained what was happening and asked him to let my boss know if he asked about me.

We were waiting for ages. Roland was telling me all about his wife who was heavily pregnant and how excited he was about the birth. After sitting for hours in the hospital we met a lady who had a look at my eyes. She put some dye in them. She said it may be a cyst and she would book me in to see herself and her boss the next morning first thing in Mayday Hospital. She said I might be putting pressure on my spine and the fluid would need to be removed. At this point, I had dark yellow dye in my eyes, and couldn't see anything. I guess I thought it was my eyes that were the problem, and thought I might lose my sight. It was frightening.

We drove back to my house and I was really tired from the whole day. I texted my parents to let them know that I may need to have an operation, but didn't go into too much detail as I didn't really have any answers and didn't want to panic them. Roland said he would come back again later so I gave him a house key. I cooked an easy dinner and then spoke to Camilla and told her what had happened. I lay in bed and prayed all night. I felt really lonely until I prayed. I felt like it was a film and I felt bad for whoever was playing me. Still praying I fell asleep.

## 2 'I'm very, very sorry'


On Wednesday 21st May, I woke up and Roland was already downstairs. I phoned my colleague again to update him in case my bosses asked about me. Roland and I took some food with us as we knew how long these things could take. We were praying about everything. We spent the whole day in waiting rooms laughing, even though we suspected that we were not going to hear good news. God's strength kept me very positive for the whole day, and if I shut my eyes, Jesus was there smiling and saying "Don't worry." I saw three different doctors, including the lady from the night before. They were prodding and looking at my eyes. They still thought I had a cyst and they wanted me to have an MRI scan later that day.


First, I had to go to one other doctor to run some more tests. At that point I wasn't really with it. Roland just took care of everything. I do remember the doctor had a bit of an odour problem, no offence, but he was very, very smelly, and he looked like one of those old doctors who may love their job but doesn't particularly like people. We had to try hard not to gag. The doctor was looking at my eyes and said he would get an MRI scan set up and some blood tests too. At that stage, I thought that I might have to wait for an hour, go home and come back the next day to look over what I needed to do. Roland lived close to the Mayday Hospital and he went home to get some stuff done as I was going to be waiting for a long time. It also meant that he could park my car at his house instead of paying for the hospital car park. I sat there for two hours, during which time, I texted Camilla and Becca. It was feeling very surreal, like this was happening to someone else, and I wasn't sure what this cyst would mean. I just kept praying and kept positive.

Eventually I had the MRI scan. It was very daunting as the camera was an inch away from my face. I shut my eyes – it was so loud. Tears were streaming down my cheeks and for the first time that day I was scared, really scared. I suddenly heard a voice saying "I'm here." I couldn't move but I knew God was there. He smiled at me and said: "Not very nice in here is it?" He stayed with me the whole time. I felt so protected. I knew I needed to remember that he was there all the time. The crazy thing was, that day I was not in any pain, it was just all very confusing, but I knew I wasn't going through it alone.

I left the scan room and went to have some blood tests. I was just calling Becca and Camilla to update them when a lady ran over and said "Hi, is it Claire Bullimore?" I said "Yes." She took me back to the part of the hospital I had been to before the scans and I waited to be called in. There was an old lady waiting as well as an old man and a couple. The older lady was offering everyone sweets and it was making me laugh talking about her piles, but I soon lost my appetite.

An hour later it was just me and the couple waiting. I could see they were Christians; we talked about our Churches and about God. The wife went to a Pentecostal church. Her husband went to a Church of England church. He was saying he didn't like all the singing and jumping around at a Pentecostal church. I said I love singing and dancing so it was perfect for me. Then they went into the doctor. So, after waiting two hours I was the last person. I truly believe God let me meet some great characters in the waiting room to stop me sitting there and worrying.

The doctor finally called me. I was relieved that they must have found something rather than sending me off with nothing. The doctor was making small talk about my job and my school history. I was just thinking, "Please get to the point." I could see the X-ray, and it was dark but full of shapes everywhere that looked like a cauliflower. I had no idea what it meant. Finally I could see he was getting to the point.

He smiled at me and said "I'm very, very sorry but we can't help you, you have a very small percentage chance of recovery." He kept saying "I'm so sorry" all the time. I was just thinking, "Yeah sure you are." He showed me the X-ray. "This is the tumour," he said. "Tumour? Will I die?" I asked. I could see what he was referring to, it was huge and all over the place. He said it was so big that it could kill me any day if it was not removed or treated.

I just sat there looking around the room trying to get my head around it. "A tumour," I said. He said he was surprised I had not been showing symptoms a long time ago as it was one of the biggest he had seen. Must have been growing for around 10 years he said. I said I had been telling my GP I had problems for years but they wouldn't help me He said he was going to get me a bed ASAP and get me into surgery. I felt sick. I just said "Wow, OK." What else can you say to that?

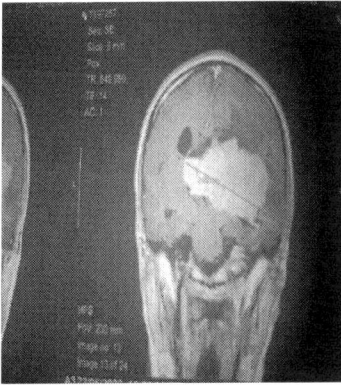

This was my tumour. It was the size of a grapefruit (10cm). The black dots were fluid that was building up.

The doctor said that he would go off and organise a hospital bed at St George's for that night and asked if I wanted to stay with him while he was getting some things sorted. I said "No thank you," and that I was fine. I refused to cry in front of him or anyone else for that matter. I just wanted him to go away so I could take in what he had said. The minute he walked away, I thought I was going to throw up. He went away for about 20 minutes which felt like an hour. I just remember thinking "Is this for real? I'll never see my family and friends. I'll never have children or grow old with someone."

My parents and Ian were in Australia and might not make it back to see me. How was I going to tell them on the phone? I went numb all over. I prayed for the strength to handle it, and I really hoped that I would go to heaven as I would see people again in the end, and that gave me some comfort. I wiped my eyes and just accepted it, and I wanted to make my last few days good ones with a smile on my face. I called Roland and told him what the doctor had said, and that I needed to go to St George's Hospital later. He said to come straight over to his house, and he would take me.

The doctor came back and gave me the details for my admission to St George's. He held my hand and said really slowly "Go and do what you need to do and then come back later, they'll be expecting you." I could see the doctor was not giving me any hope. I didn't really know what to say to him except "thank you for your help" I laughed at this as it was somewhat ironic.

It was 16:30 by the time I left the hospital. I walked to Roland's mum's house five minutes away from the hospital and rang the door bell; I'd never known him get to the door so quickly. He gave me a huge hug and I burst into tears. I think I always deal with things OK on my own but the second I see someone's face, I'm off again with the waterworks. When he hugged me it really made me realise the seriousness of what was happening. Roland looked at me and said that he thought that they had made a mistake, he said he couldn't explain why, but that the doctor had made a mistake. "It is not your time. God is in control," he said quietly, and I quickly agreed with him.

I went in and sat down. I needed to make some calls to see if people could see me before my operation. Although I knew that Camilla, Becca and Emma wouldn't be able to make it I phoned them first. I wasn't sure how I was going to tell them. I called Camilla first as she had been calling me all-day and keeping in touch. It broke my heart to tell her the news. I remember her crying and her voice was so sad. I tried everything not to let her panic and to be very positive. Her mum had to take the phone off her because she broke down in a shop in Reading and I explained to Sylvia what had happened.

This was another time I knew God was with me because I wanted to cry so much, but I had God holding me up and I was completely calm. I called Becca but she was so relaxed and I couldn't believe how easy it was to tell her. Later I found out Roland had called her already with a heads up so that she didn't get upset on the phone. She put on a pretty amazing voice and I was shocked at how easy she made it. I called Emma, but it went to answer phone and I didn't think it was appropriate to leave a message. I then called my parents in Australia, and Mum answered the phone. I tried to be really to the point and I kept being positive, telling her God was in control. I knew they didn't believe in him but he really was with me. Mum said they'd get back to the UK as soon as they could.

Mum said that I should call her sister, Anne, and she could speak to the hospital as she used to be a nurse and could understand all the hospital jargon. I did, and Anne said that she would be in the hospital first thing the next morning. I had the whole day in hospital on the 22nd May 2008 and my operation would be on the next day at 08:00. After finishing on the phone, Roland said that he had spoken to our Pastor and Sister Gill and to our friends Everton and Vanessa, and that we would all meet at our Church to pray about the situation and for a miracle. First we went to my house and met Vanessa and Everton there. They were in shock, but I just said we must not worry. We met Pastor and Sister Gill at the Church. They all placed their hands on me and we asked for a miracle and that His will would be done.

Roland, Everton, Ness and I went to get some takeaway from a Chinese. I went home and collected all my things. I had the best friend I could ever have with Roland; every time I would slightly doubt, he just made a joke to lighten the situation and I would just laugh with him. I tidied the house up completely and emptied all the bins. I think that was my obsessive-compulsive disorder (OCD) coming out. We went back to the hospital at 23:30 – we were very late but I decided that if I was going to be having hospital food for a long time I was getting my Chinese in first. We went to the hospital and I remember my vision was getting bad again. I think the body can give up even if your heart hasn't. The night nurse said they were waiting for me. I knew that I was going to be there for a long time, so I didn't hurry but I did apologise. I said goodbye to Everton, Ness and Roland and I got into my hospital bed. Surprisingly, I fell asleep very quickly.

I woke up early the next day hoping that my parents would get there before I went down into theatre. I phoned my work and explained what was happening to HR and spoke to my friend Nirm to explain. She was really calm, even though it must have been a big shock. Nirm said that she would let everyone know and organise a card, she would start a collection and I said perhaps I would have to use the money for some nice wigs, as I was told my hair would be shaved.

Avoiding talk of the spectre of the worse outcome, we talked about when I would get out. We said we would go shopping when I came home and buy some funny wigs, perhaps blue or red. I never thought that I would have to ask for a wig as a present. I noticed that I had loads of missed calls from people at work, but I couldn't deal with talking to them. I would just let Nirm pass on the news. She really couldn't understand how I could be so calm. I explained that my faith was strong and that was all. I also phoned Camilla and Becca to talk to before the operation.

My Pastor came in with Sister Gill to see me and brought me a card called 'Footprints in the Sand'. I really love the quote and I then heard Leona Lewis had brought a song out that was called 'Footprints in the Sand'.. It was a song that just felt like it was written to encourage me, which it did.. It was overwhelming how many people were all there for me. Aunty Anne came to see me, and went to speak with the doctors to get a better idea of all the details. My cousins all came in to say hello or just called me. People were so nice, but I could tell they didn't know what to say. I had never seen so many magazines in one place. I just smiled all the time because I was very content and I felt that I knew I would be fine. I just wished everyone could have been as confident as I was. My parents called me in the late afternoon and said they had got to Kuala Lumpur and were hoping to get to me first thing Friday morning.

My surgeon, Dr Minhas, came to introduce himself to me and explained what was happening. He discussed potential risks, such as problems with balance, coordination, speech, vision, memory or muscle function, depending on which area of my brain was being operated upon. He said he would not know the extent until we were in surgery. I asked him not to take too much of my hair off if he could help it. I had just had my hair done, over the weekend. I laughed, but he didn't.

I said I had already sorted out a wig for later on. He still didn't laugh. He said I had a very, very large tumour the size of a grapefruit and it would take two operations. He said it would take around seven to eight hours for each one and I would have to have the second operation later on. One down side was that I might lose my vision as the tumour was in a tricky place. I said "I trust you. God is in control and he already told me things will be fine." He smiled with a doctor's smile, like "God? Oh yes, hmmm!!" I just grinned. I was totally confident. I didn't even spend time thinking what would happen if I lost my sight or speech. Everyone left my bedside and I used the time to call Emma. I called and told her the situation and not to worry. She was really strong for me and said I would see her after I had the operation. I know that she was very worried and later on she told me how she broke down in tears. Luckily she was staying at the house of a friend who was able to console her.

I woke up very early on Friday 23rd May 2008 and the doctors came in at 06:00 to give me the especially unattractive hospital nightgown and to fit me with my arm bracelets. It was all becoming very real and I continued to pray for strength and that I would not be worried, as no matter what - I would be fine. I filled out all the paper work for the operation and they said they would be collecting me at 08:00.

20 minutes before I was due to go down for the surgery, the door swung open and my parents came running into my room. "We're here!" They had come straight from Heathrow Airport. It really looked like a scene from a movie. Sort of went into slow-motion. I could see they were relieved and so was I.

I told them St George's Hospital is one of the best neurosurgery units in the country and I hoped that fact would give some comfort to my family. I could see that they were worried and that made it very hard for me. Two orderlies came to take me down to the theatre. I was still trying to be strong but then I saw my dad was crying. I knew that my family and friends had no idea whether or not I would even come back. I waved goodbye, smiled and said "I'll see you soon." The surgical team were standing over my trolley now, and talking to me. The next thing...I was out like a light.

OK! Go and grab a coffee or a toilet break. That was intense!!

## 3. 10 Hours Later

I was in surgery for 10 hrs. The hospital had sent my parents home and they just waited by the phone for the hospital to say the operation was over. The phone never stopped ringing with family and friends; everyone wanted to know if there was any news. It must have been terrifying just waiting and waiting. They eventually gave up waiting and came up to the hospital because they were afraid of falling asleep and missing the call to say how it went. 10 hours later my parents were told the operation seemed to have gone well.

Now it was a waiting game until I woke up and they could assess the effects of the surgery. I was kept asleep in the intensive care unit for a further 24 hours. I know my parents and Becca came in every day, but I was never awake much. I would drift in and out of sleep. After three days I was taken into the high dependency unit for two weeks and then went to the normal ward for two days until I left on the 6th of June.

Fantastic news came from Dr Minhas. He told my parents he was able to remove the entire tumour in one go; I had an intraventricular meningioma. Only 2% of people ever get this type of tumour. I wouldn't need another operation! Dr Minhas told my parents the tumour was so big that when they removed it my entire brain moved over. I didn't know the good news yet myself but I was not worried at all. I knew God was in control and no matter what happened I would be ok. We would have to wait for the tests to come back, but Dr Minhas was sure the tumour was benign.

Intraventricular meningiomas are rare tumours. The origin of these tumours can be traced to embryological invagination of arachnoid cells into the choroid plexus. What a mouthful that is lol.

Intraventricular meningiomas are slow-growing tumours that can grow very large prior to detection. Although they are commonly seen in the lateral ventricles, they occur in the third and fourth ventricles as well. Intraventricular tumours cause headaches for some time as much as a recorded 16 years prior to detection of the tumour. Headaches, intermittent or persistent, are the most common symptom.

The most common effect of surgery to remove an intraventricular meningioma is speech disturbance. Recovery can be difficult if there is a postoperative cerebrospinal fluid leak. Most people with this tumour have visual field deficit, disconnection syndrome, or speech and cognitive deficits.

A diagnosis must be established using MR imaging, and surgery requires planning to avoid eloquent area damage. Early control of the vascular supply to the tumour is critically important, and the tumour can usually be removed intact without damage to vital areas around the ventricles. No recurrences have been reported after complete excision.

That was what the doctors said but everyone is different so it would just take time to see if I had disabilities. Whilst I was in the land of nod. I was having some really crazy dreams. I never knew if they were real or not. I had to ask my parents or Becca when I woke up.

I was only scared once with the dreams and that was at the beginning at the operation I think, as I dreamt the anaesthetic didn't work and I hadn't fallen asleep yet and no one had realised. I was going to be awake during the operation and I was trying everything to tell someone, but I couldn't speak or move. It was a room full of other patients waiting to be operated, one after the other like a car wash because once you were locked in you couldn't move until after the process. I did panic a bit and I was asking God to help me to relax. This was not for long, but it was quite horrifying. I then went to a completely different dream where I was feeling God's presence and that I had nothing to worry about because He was there the whole time just watching over me.

After the strange dream about the operation, I could see someone, who was similar to my uncle John, who came every time I was sleeping and stood with me no matter where I was. He never spoke, but I knew that God was sending a messenger to let me know He was there. At one time, I could see what looked like ghosts, the kind you see in the children's programmes lol. There were these floating ghost looking images all tied to crosses like the crucifixion. I just smiled because I knew that the Devil would love to use this kind of opportunity to scare me, as I was helpless. I am such a wimp normally, but I felt very comforted knowing God was with me.

In one shocking dream, I was sure that I had woken up during my operation; my body was like a very heavy weight. I had no feeling; all I could do was move my eyes. It was strange, but it didn't hurt, it was just a little unnerving. I heard noises everywhere. To my surprise, I was told after leaving hospital that I had been woken up in the operation and that was a normal check on brain function and a test to see if I could say 'yes' or 'no', or just blink and look at them. I fell asleep very quickly, but it was a very strange few minutes.


In the high dependency unit, I slept most of the time, only coming round now and then. I couldn't speak, but at that point I thought that was because I was just tired. However, I was about to find out that my speech would be affected for a lot longer than a few days. I could understand what I was seeing, but couldn't understand what people were saying to me. I tried to speak when I was asked questions, but it didn't come out right. The doctors explained that I knew what I was looking at, but when I tried to speak it was just jumbled up. They said my speech was going to be hard to recover and they wouldn't know how much I had lost; only time would tell. I just smiled because even though I was talking rubbish then, I knew that God hadn't brought me that far to deny me a full recovery. I was just so certain I wouldn't accept my disabilities.


I was very happy to see people even with my jumbled or non-existent speech. Becca would be there every day and sit with me and I was so glad to see her. She would try to brush my hair a bit as I was looking like the Grinch. She would cut my fingernails as I was scratching my face without realising a lot. I was blessed to have such a great friend.

I remember Roland came in with Reka one day. They were flying to Hungary and came to say goodbye even though it wasn't visiting time. Roland was being his usual self, getting around the nurses. I smiled a lot and hoped people would know how thankful I was for them being there. I was more than happy for people to come and say hello. I would continue to smile and watch everyone else talking to each other. I didn't have a clue what they were saying. I would smile and then fall asleep. I remember my cousin Clive coming in and I smiled and fell back to sleep. I wasn't the greatest conversationalist to be honest, hee hee!

It seems I had got into a bad habit of sleeping all day and staying awake all night. I remember the nurses coming round in the evening to wash me and I was so out of it. I was embarrassed but had no coordination to say I didn't like it. I couldn't speak yet, let alone move properly. I had one night nurse who would give me a number of my medications. I had to take around 20 tablets, which were spread out over the day. I hated the night ones at around midnight. I didn't realise this was procedure, and I was convinced my night nurse wanted to kill me lol as she was very heavy handed. I remember thinking, "Step back lady, I'm a child of God. Ha ha!!" All the nurses were lovely. I had to have injections all the time and help, because I couldn't do things myself yet. All the needles hurt but they were not horrific, and by the last week I would just look away and say "Yeah, fine. Go for it – use my bum, if you like, as the other parts are looking very bruised and I want to wear short sleeved tops again."

My arms were looking like pincushions and were multi-coloured. My blood was taken every day, and when one of the nurses came in, I asked, "Will you be leaving some of that for me?" She never seemed to find my sense of humour funny. I can't imagine why not. I had another man from Poland who was giving me my antibiotics and I asked him, "Can you change my arm as that one is a little sore?" He always looked really guilty, but I would just give a big closed lip smile and take the needles. The one for the drips were the worst.

One of the down sides to having brain surgery is seizures and I unfortunately had my first one in the hospital. It was not ideal but this will make you laugh, or cringe…. After the surgery I had a head drain which was there to basically drain the fluid that had built up where the tumour had been removed. I was always told I had to have the drain above my head so the fluid would go out properly. I couldn't get out of bed until that was out. After a week I had the drain taken out and I could sit up and have the Catheter bag taken off too and finally stand up after a week.

This one was embarrassing but I can laugh now, kind of, anyway. I had three drips in my arm and I thought one of them was my food as I was never hungry. I was told they were worried I hadn't been eating and they wanted me to try. I looked at my three drip bags and I said "Hmmm, those three bags are my food and medication going in to keep everything in liquid form right?" "No, they are just medication; you need to eat more to gain your strength." I'm like, "Oh boy!!!"

I soon needed the loo after a few minutes. They gave me a bucket on the bed and I was like, 'No way, that is messy and just really not nice lol', I wanted to try but I just couldn't as I was really not liking the idea. The nurse then pulled over a portable toilet. I know I hadn't eaten much but I still seemed to need a number 2. (Yes, that's what I said, stop laughing boys and girls, stop hiding your faces behind your hands!).

I tried to get out of bed and stand up and tried to sit on the portable toilet, I didn't realise how weak I was and my body was like a dead weight and my arms were shaking. Unfortunately, after doing my business, I had a fit. I opened my eyes and I was back in bed with a gas mask on my face feeling very sick and surrounded by nurses. I felt better after a few minutes. I wasn't sure what had happened, but my mum said I had had a fit.

Oh great I thought 'Hello I am 'THE POO GIRL WHO FITTED'. Oh the shame. It's just as well I have a good sense of humour. I was still determined to get up on my feet, and go to the normal toilet..

Slowly I was wobbling around the ward and leaning on the handles that were all around the room. I was very weak on my right side of the body and was told I would need physio to help with that and sensation and feeling was not really there.

My thought processes were definitely still interesting; the ward would not allow too many people to come in. I would call people by the wrong name or just didn't know their names at all. It would tire me out to talk as it took me a long time to say what I wanted to. I started to have more family visitors and tried hard to think and speak. I told my grandparents who are both very short, more like 5'4"ish, that they would never have the same problem that my feet were having. They were always sticking out the bed. I had to wear lots of socks to make up for the short mattress and they seemd to feel really cold all the time and never warmed up. They do give you the socks in hospital but for some reason one of the nurses rolled mine down to my ankle and although I didn't feel it until afterwards, getting out the hospital the hospital socks have left a scar on both ankles where it cut my blood circulation. I see them as a reminder that I'm still walking and alive however it does look a bit weird. Like I have some bandage on lol.

My parents and my grandparents and Becca had come in every day. I asked my mum what the history was behind my nickname Wookie, which my grandad kept calling me. She said my uncle Chris (one of my Godfathers) was a Star Wars fanatic and didn't like my dad's nickname for me, Clairey Cluey. Chris said "Hmm, 'Cluey' rhymes with 'Chewy', which is Chewbacca's nickname in Star Wars. Chewy is a Wookie Bear...I think you should be called Wookie," and the name stuck. I was, like cool!! I would call my dad "one of my parents", as I couldn't remember the right word and with Emma I would just smile a lot because I really didn't know her name.

With some prompting I would just about understand who people were but I couldn't really remember much about them. My mum said I had something like 30 texts from people on my mobile phone while I was in hospital. I couldn't read them at first because I couldn't read anything and had to wait for my mum to read them for me. My Pastor and Sister Gill came in to see me again. They sat and prayed with me. They were so supportive.

I was able to watch the TV or listen to some music finally. Hospitals have gone up in the world with electronics. You can do anything from your bed without asking for help. You have to pay for the TV, but it was a small price to pay instead of hearing what everyone is doing around you in the ward. The ward itself is like watching Casualty, but it gets depressing after a while, watching sick people.

I had stayed away from mirrors initially because I couldn't walk to the bathroom and then because, even though I could, I didn't want to see my scars yet. I knew some of my hair had been shaved off on one side and I was really nervous that it would be obvious to people. Finally about four days before leaving the hospital I was able to have a bath on my own. I had a nurse helping me. I couldn't get water on my scar as it needed to stay dry. It was still the nicest bath I had ever had. I didn't want to get out. I hated the whole being bathed in my bed every day by different nurses. On the 2nd of June I was moved to another room. It was nice being taken out of the high dependency unit and to have all machines taken away. . Now I was just waiting to be sent home.

I was sharing a ward with six other ladies now. They were all there for different head operations. One lady opposite me was in for the same reason as myself. She was Greek and very loud but lovely. Her family would come and see her all the time. Another lady was next to me and was there for a tumor that they could not help her with. She would die at some point within a year she was told and she was only about 50. She was such a nice lady but I could see she was scared. I would just pray for her a lot that she would not suffer.

## 4. Time to Go Home

On the 6th of June the doctor came in to say I could go home. He gave me a very large box full of tablets. I looked like a walking chemist. He explained a few things to my parents and me. I wasn't really paying attention and I didn't really take in what they had said. Looking back now, it was information I wish I'd listened to. Then it wouldn't have been such a shock later on, having to wait for the rehabilitation.

The doctor said I would have short and long term effects. He said that swelling of the brain can cause weakness, poor balance and coordination, personality changes, speech problems and fits. He said the symptoms would usually lessen and disappear as you recover. He said that this process could take months after such an extensive operation. Because of the position of the tumour, I might have long-term problems with speech or with weakness of an arm or leg. It might be hard for me to keep my spirits up through this time. But with effort and help from physiotherapists, speech therapists and other rehabilitation specialists, the doctor was confident that I would get a lot better.

My rehabilitation would start as soon as I could get out of bed alone. I was told not to expect instant results! I would gradually be able to do more and more for myself. I was warned that I might never quite recover the same level of fitness as before my illness, but that my condition would improve. I was promised that my confidence would increase as I learnt to manage with whatever level of disability I had to cope with. Brain tumour treatment can feel like a long haul. I knew that it would leave me physically and emotionally drained. I would need time to recover my energy.

I was told I couldn't drive until a year after any seizures. So now it was a waiting game. I had to hand my license over to the DVLA and wait until we could confirm that nothing else had happened seizure wise. Luckily I don't have to re-take my test again if and when I can get back behind the wheel.

I remember standing up all ready to go home. It was so nice to wear my shoes again although I kept say "I have my feet on" instead of shoes. I brushed my hair very carefully and tried not to get upset as I was finally going home. 17 days ago there had been a question over my life and I just kept holding onto God the whole time. He truly is amazing to me. I wobbled out the hospital. My legs were like heavy boulders – they felt as though I'd just run a marathon. It didn't matter how much I wanted to walk faster out the door, I couldn't. I kept looking at my hands in amazement because they had finally stopped shaking after two years. I was very spaced out but I was happy to breathe fresh air rather than the hospital wards. I put my thoughts and situation into God's hands and was very strong in hospital, but when we were driving out of the car park it dawned on me that people didn't think I was coming back. I realized how much God took care of my emotions in hospital.


As I got home I started feeling very sick, the kind of feeling that sits in the pit of your tummy. I was determined not to cry, because I was happy, but I was really emotional. It was very overwhelming how much mail I had when I walked into the house. I had told my parents not to bring me any cards or presents as I didn't want to get emotional, and I was on a mission to get up and out.. I could tell from messages in the cards people were very worried about how extreme my situation was and that was something I hadn't wanted to see before but afterwards, it was very very humbling how caring people are…. My company had sent the biggest bunch of flowers I had ever seen and the vase was gorgeous. I went upstairs, very slowly I might add as I was very weak and my balance was off. I went to my room and it was very strange being there, even though everything was exactly as I left it. It was a very similar feeling I would get when I went home for the holidays from boarding school. When I returned it all felt new for five minutes before the old familiarity set in.

I looked around; everything was as I had left it. I read through all my cards and people were so lovely, even people I hardly knew had sent cards. One of my best presents was that my company said I had been made permanent two weeks prior to my getting back home. I was so happy. I just felt so grateful and it was a huge weight off my shoulders as I had finished my probation a few days before I went into hospital. I thanked God for the news. It felt like everything was amazing, like God had put his hand on me, and everything was perfect. I don't mean the situation but my mind-set was strong.


My mum said I was going to be off work getting better for some time and I should find things to do. I felt like I had let people down, especially work, which was hard to deal with, I thought I would be off for a few weeks but, I never imagined I would be off for six months. Most people would say how cool to have that time off but I didn't know how not to work. I couldn't do much for many months and could only relax. It seemed like the next part in my life was going to be harder still. I felt I'd won the battle but still needed to win the war.

The minute I got home I called Camilla who I hadn't spoken to since just before my operation. I tried to make some witty comment but I found speaking really hard and nothing would come out right. She wasn't around so I left a jumbled up voicemail to say, "Hi guess who." She called back in about two minutes. My speech was like a game of charades. It was difficult, so I let Camilla do all the talking which made a change as I am usually the one with the motor mouth. Any phone calls were hard as I couldn't use my hands for expressions and mannerisms. I really couldn't understand what people were saying. I could see their lips moving but that was all. It was like people were speaking another language. I did get caught out saying some rude words I would never normally say. Camilla and Becca would just laugh about it with me. I avoided everyone except the girls and I never picked the house phone up – I let my parents do it.

When I looked in the mirror for the first time in a long time I noticed my pupils were really large and you couldn't see the iris in my eyes. It looked really weird. The doctor said my pupils would get back to normal after a week or two.

One of the first things I did back home was to have a nice hot shower with some privacy, not being watched by nurses' eagle eyes. Just to be sure I didn't fall over Mum stayed nearby. I was told not to put the shower directly on my scar for about two months. It was so nice when the doctor said I would be fine now having the water over the scar. I loved having my full showers again.

Camilla rushed down to see me on the second weekend. It was really nice to see her. We just laughed and caught up on the gossip. Well, her gossip as I was severely lacking material in that department. She bought me a lovely tape she had made for me. She also bought a lovely silver necklace with a cross. It was really nice talking again. We were both trying not to get emotional over what had happened. We just laughed at things. I couldn't talk too much about it as it was hard. I was very pleased to see her.

Each day I thank God for my family and friends for putting up with me and I will never forget the people who stuck by me. Thankfully my work company kept in touch with me and was very supportive also.

Mum and Dad offered to help me paint my room to give me something to do. Becca helped me to move the furniture around, or should I say Becca and my parents moved the furniture around and painted as I still wasn't allowed to pick things up. I had to be very careful not to allow the build up of fluid in my head to get any worse than it already was. I could never stand up too long for months after the operation. I was aware that my sight on the right side wasn't great. I can see everything except at the corner of my right eye. It was crazy because I was still not speaking well and I wasn't ready to see too many people but the thought that I may not be able to drive ever again worried me a lot. I was told it may get better and I was going to have an eye test at some point when I hoped the doctor would say I could drive again.

In July 2009 my next door neighbour had a bonfire on a very windy day. The fire set light to his pigeon loft (no pigeons in the loft, don't worry) and was quickly out of control. Our very good family friends Brian and Beryl were staying with us for the weekend and on Sunday morning as we sat outside in the garden we saw what was happening next door. Everything turned out fine; two fire-engines and an ambulance arrived, and the fire was put out. Just a normal weekend down my road…no really, it is a normal weekend! It was the funniest thing, I was trying to run for the first time to look at the fire but it was impossible for me to move quickly. Ha Ha!!!

I only went out very, very occasionally for the first two months. Becca would pick me up and we would relax at her house. I was always too tired to go far or for very long. One other thing that was amusing back then was I always left my flies down even when I was putting my jeans on in the morning. Then later in the day I'd go to the loo and realise they were open. I'd see what I'd done and laugh and then forget to do them up again! There were and still are a few mind blocks, but that one I sorted out pretty quickly.

I was very thankful to be home but things were different and I was finding it hard to be patient. My good friend Naomi from college came over on the second day and said she just wanted to say hi. What she didn't tell me on the phone was that a van was going to come that day and bring me a huge bunch of helium balloons saying get well soon. I had never seen so many balloons in one place, it was such a nice surprise and I was very happy to make use of them when they were going down. I could suck out the helium and have conversations on the phone with friends. It's the little things that made me giggle; OK, OK, I know, but give me a break – I didn't get out much!

Reka and Chanel kept in touch all the time and came to see me. I really am lucky to have such great friends. Now I was remembering things every day. I could remember so much more from my childhood and all the years in between. I truly had no idea how much of my memory was gone. Unfortunately, before the operation I would have a conversation with someone and a moment later I would not remember what we had said. I knew faces really well but any other memory was gone. In the few months before I went to hospital I was very cloudy about what had happened but, my friends just filled me in and I would eventually remember what they were referring too. I had lots of pictures in my head but I did need some help with putting them into words.

## 5. Sent to Rehab.

In mid-July 2008 I started with a private speech tutor who helped me for three weeks, getting me ready for attending the NHS Neuro-rehabilitation centre in Croydon. She explained what had happened to my speech and that I should be able to have most of it back but it depended on how affected the brain had been. It may never be perfect but with strategies I could cope. She said that usually words and memory are all stored, in the brain, like a filing system. When we want to remember something or speak, we automatically go to the right file to get it. During the operation all these files had been thrown up in the air and had become totally disorganised, with the result that it takes longer to find the right file.

After three weeks I was finally able to start my rehabilitation programme for speech and language therapy, physiotherapy, occupational therapy and relaxation therapy. The relaxation and occupational therapy were to help me cope with everyday things and get me ready to return to work. The first thing I thought when I heard the word 'rehab.' was Amy Winehouse or Britney Spears. I think most people relate rehab to people with bad habits, or at least I did.

When I arrived it was difficult, it felt unfamiliar and not very nice, especially as I wasn't really over having been in hospital. It was the first time I would be talking to people other than friends and family. I remember that on the first day I went to meet my new speech and language teacher Michelle. She seemed really nice and was very welcoming. I had to answer a bunch of questions about myself and what had changed for me after the operation. She said, right at the beginning, "Please don't worry if you cry at this part, as we understand it is a very difficult subject to talk about." I just smiled and said "No, no, that's fine, I'm fine with everything." Within minutes I had burst into tears lol. There went my chilled-out look, right out the window. It brought everything to my attention, about what had happened and how extreme my situation was. It was highlighting all the disabilities I now had. That was only the first day at rehab and I wasn't comfortable with things at all!

I then had my physiotherapy with Elaine the next day. She was really nice but very firm which I liked because I knew she would get my body sorted out. She had to do an assessment to see what parts were causing a problem. The assessment showed I was very, very weak on my right side and that my balance was really bad. Because I had lost vision out of the right corner of the eye it took me a long time to adjust my peripheral sight. I would get really tired very easily. Elaine would say it wasn't about doing things for hours, it was about doing a small bit each day because I needed to take it easy. My rehab classes were all about long sessions with lots of breaks in between. Elaine felt I would benefit from doing Pilates. It's a great fitness programme that you can push yourself at, at any level. Anyone, at any age could do this. It all depends on your own levels of fitness.

My next session was occupational therapy. I met Susie and she explained that she would be the person to talk with my HR department to plan a back-to-work strategy. I then had to do some questionnaires to see where my weaknesses were. She asked me what my most important goal at rehabilitation was. I said I wanted to get back to work as soon as I could and wanted to get back to my normal life. She was very happy with my determination but she said I needed to understand I would be in the rehab for three months and I would not be able to rush back to my old life. That was another very hard conversation. I felt helpless but I was determined to get back to my life and I knew God had already brought about one miracle in my life. In Matthew 19:26 it says "with man this is impossible, but with God all things are possible". I just kept this with me even when it seemed pointless. I wasn't going to give up. It was very easy to put my head down and cry, wondering why did this happen to me? I did a few times but, no matter how much I would get upset, I would just keep praying and reminding myself that everyone is on this earth for a reason. It was up to me finding out what that reason is and to be the best at it. God had done his part for me, now it was my turn to do mine!

It was in rehabilitation that I decided to start turning my diary into a book. It was a way of keeping me busy. I didn't want to feel sorry for myself. I needed to keep positive and the book would remind me that I had come a long way and I wasn't about to give up now.

I started relaxation classes in the last week at rehab. I was always the youngest person in rehab but, I met many great people who were mainly there after having a stroke.

I did meet one lady and she had the same tumour as I did, except when she had her operation they had to go through a different part of the brain. They have to make a choice about where they go in to get the tumour out. I knew that was a choice they had had to make during my operation and that the choice was made to give the best chance of saving my abilities, both physically and mentally.

With the other lady they went through a different part of her brain. She had lost her movement in a very large way and she was always in a lot of pain as she was very weak. I was operated in the left side of my brain and I lost part of the sight in my right eye and my movement and touch were weakened but it was mainly my sight. I had lost a lot of my speech too.

I was exercising a lot to work on the strength in my right side. One of the other things I was doing, to help my speech and memory, was my singing. I have two hobbies in my life and one is singing but my speech had been hit hard by my operation. I couldn't sing anymore. I could hum the tune and read the words in my head but when I opened my mouth nothing came out. It was like a footballer being told he may never play again after breaking his leg. Even though he knows all the rules and what to do, without his leg he would be powerless. It was very frustrating but I was determined to sing again and I spent hours and hours on a karaoke as it meant I could read the words and see what I needed to sing. Day by day it seemed like nothing was changing but week by week I got stronger. After about six months I was starting to get faster.

Singing was not my only challenge, although that was one of the things I badly wanted back. I didn't remember half the things I am now recounting to you. It was horrible losing my memory. I can smile now and say my memory is nearly back to 100% but I don't think it will ever be perfect. I shocked myself when I remembered things I couldn't before my operation. It took a long time – I can't believe how much I had forgotten. I thought that because my memory had gone, it would never come back but it is amazing how all my memory is there and coming back every day.

One day I went downstairs to play on my piano and amazingly I could still play. I have never been able to read music and just play by ear. I couldn't believe I could still do it. I had been forgetting how to play the piano since 2007 – I could only play one song. I can't explain my playing to this day. I still give God the glory that with determination I as battling my obstacles.

Other things I spent my time doing during the months off work were watching the whole of the Wimbledon Tennis and the Beijing Olympics, especially as Blake was competing in the diving. I had never really watched either since Linford Christie was running the 100 metres. I had never watched the tennis but had always liked Nadal from the odd matches I had seen on TV. I was ecstatic when he won in 2008 and felt quite privileged to have been able to watch the whole tournament.

In August I had a lovely surprise when my brother came over from Sydney. He had come for Blake's homecoming party at his dad's house next door to mine.

I went to Church on the first Sunday I was home and I couldn't wait to see my pastor and the church. Unfortunately it was very hard as I couldn't sing or understand what people were saying to me. I couldn't tell what had happened in the sermon. The information just didn't go into my memory. I couldn't write quick enough either and it was very frustrating, I felt that God had been taking me on a journey that went on for ages and that was discouraging but wasn't going to let it stop me trying.

I found other people were finding it hard to understand what had happened to me. Lots of people were asking me questions and the more people I was meeting again the more I would find it very overwhelming. Small crowds were beginning to sound like a football match to me, I found it all overpowering. I think I found my faith very easy in hospital, as I had nothing to do but have faith. Once I was home I had a lot of worries that would stress me out. People would ask a lot of questions that I couldn't answer like, would the tumour come back one day? a few times I know I lost my faith in God. Not because I didn't believe in God but I never knew what was coming around the corner for me and every time I went to the hospital it seemed that I was always given something else to add to my pot.

All the medication I was taking was making me very emotional and all over the place, and my skin was disgusting from all the tablets. I was constantly very hot and my face was glowing all the time. It sounds as gross as it was I was very uncomfortable leaving the house. I just sat in the garden while the sun was out and put all my old work notes together to remind myself, what I used to do at the office. I also went through my entire photo collection. I have always spent hours taking photos and putting them together in artistically designed albums. I'm not the kind of person who can sit still for a long time. I was finding any excuse to keep my brain working. My mum bought me a DS with the Brain Training game. Just to show you how poor my memory was: it took me three months to get my score down from 80 to 30. I am not at 25 yet but I am on the training every day and from 80 to 30 is pretty great. I only need to get rid of another 10 more years now!

The rehabilitation finished eventually and I was asked to just use the techniques I had learnt and build on my strengths. I was then given an appointment for my CT scan. The first scan was strange because you normally have a scan and then end up with an illness, but I was just reminded all the time about what had happened, when what I really wanted was to forget it, put it behind me and go back to my life. I remember the scan on the 19th June was quick and my parents and Becca came with me. We left and went for a walk with Becca's dogs. I was falling asleep afterwards, as always, and after the walk I was ready for bed. Becca gave me a lovely dress that was just to say how much she cared; Camilla had done the same with the lovely silver cross necklace and a music tape she made up. I was so glad to have them.

I was called the next day, from the hospital, which was not normal, and Dr Minhas said he had a concern with my scan. He said I had some bleeding in the brain and I may need to have a drain put in. This could mean another operation. I was so scared of going back in again. I think I knew I needed prayers and I sent a text to my Pastor and Sister Gill and some other close friends from Church and asked for their prayers. I didn't go into details but I was freaking out and I couldn't control it. It's like waxing or stubbing your toe, only 10 times worse. You don't want to do it the first time and you certainly wouldn't want to do it again.

On the 23rd July I went to see Dr Minhas. Becca and my parents were with me. When I got there I went straight to the toilet. Becca came with me and I couldn't stop shaking. It was horrible and I couldn't make it stop. Becca just gave me a hug and told me to take deep breaths and relax. I tried to know God was there but I was still so nervous. Eventually I managed to calm down and splashed some water on my face. I just prayed to God I could get through the meeting and could crack-up afterwards. We went in and Dr Minhas was lovely as always. He said he wasn't going to go on too much, but the scan had showed I had some bleeding in the brain and he was hoping some steroids would stop the bleeding. If not I would have an operation to drain the blood and the fluid. I left the room and I was not sure how to feel. I was certain God was in control and I just needed to relax. I was exhausted when we got home, I was numb all over. I just asked for more prayers from my Church.


I started the two-week course of steroids in the morning and over the next two weeks I became a recluse at home, as though it was some sort of cocoon that would keep any harm from befalling me. I just went to rehabilitation for a few hours every day and then came straight home. I didn't want to speak to anyone, as I knew I was getting really aggressive again. I didn't want to behave that way so I just kept to myself; I was crying at the smallest thing and couldn't control it. My teachers said they could see a major shift in my behaviour and demeanour and said I should check with my doctor. I went to the doctor, who said that mood swings were very normal with the steroids I was on. I could be happy or sad. I was not in the greatest place but my doctor said I would be finished soon. Once I was off the steroids I went back to myself. It was crazy to think some tablets that help, radically change your personality. It was very surreal and it was so great to know the headaches weren't coming back.

Good news came on the 5th of September as after another CT Scan, Dr Minhas called me and said that the blood had gone and the steroids had worked. I was to come back in November as there was still some fluid although the bleeding had gone, which was great. I went again on the 2nd November for another scan. I didn't realise that I was going to have an MRI which was very scary. I saw all the same people I had seen there six months earlier and that was quite hard. I was not fazed by the scans, I was so used to them. I just didn't realise I was going to have an MRI. I was happy to leave that day. I called Dr Minhas, to see if the results were good. I asked if I could travel now and he said YES, he said that the fluid had all but gone now and he would be writing to my GP to reduce my medication. He would send a letter out with an appointment to see him in December.

I was so happy with the news and called some friends. Emma was back in London, she had recently moved to Dubai to live with her fiancé George. We were really happy with the good news from Dr Minhas and she suggested I should go back with her to Dubai for a week. I said "Great!" and we booked a ticket the next day for a week of relaxation and getting me out of the house and back into life!

We arrived at Terminal 5 and went to the First Class area because Emma is a corporate cardholder. I was very happy to take advantage of all the free food, magazines, drinks and more food. We were so relaxed we didn't realise that the plane was about to leave. We had to rush through the terminal to our departure gate. We got there very out of breath. The ladies there gave us a very disapproving look, telling us severely that they were just about to take our bags back off the plane. Then Emma did the funniest thing and any nervous feelings I had about flying and the cabin pressure on the plane disappeared. She hobbled up to the counter and said "Ow my leg, owww!" I was trying so hard not to laugh. She continued to wobble on to the plane with noises like "Ouch!" and "Ooh". We got on the plane and could see everyone else was already seated. We looked at each other and just burst out laughing. The air hostess was clearly trying not to laugh with us and said "No problem, girls." We flew off and I spent so much time laughing and catching my breath that I didn't have a second to worry about flying or how the pressure in the cabin might affect me. We had a really relaxing week and I was so happy I had gone there. I felt like my life was coming back and I was looking forward to Christmas and New Year's Eve and seeing my friends and family.

Christmas and New Year's Eve came and went. I had caught a really nasty flu and I was really sick. I couldn't hold any food down. I was determined not to let it stop me going back to work. I went to the hospital for a check-up. I had to see a different surgeon instead of Dr Minhas. I was trying to stay positive but the new surgeon was so unhelpful. I was expecting to be told I could come off the seizure tablets and that I would be able to drive again. I didn't expect him to say I was going to be on my tablets for up to a year more! I couldn't understand why – that had not been explained to me back in May 2008. The surgeon was so happy about the job they had done and less concerned, it seemed, with my getting back to normal life.


I was still very nervous about my sight and about knowing if I would be able to drive again. I left the meeting with my parents with no clear idea of what he was saying and where it left me. It had been a very long wait but I finally went to get my eyes tested at the hospital the next day. I was told I was on the borderline of having enough vision to drive again and getting my licence back. I would need to have the test again in May 2010. Suddenly things were becoming disappointing. I prepared myself for more disappointment and once again put my trust in God.

The diagram below shows the missing part of my vision with both eyes open. This was the first eye test I had a few months after the surgery in January 2009. I had to wait a year to see if there had been enough improvement for me to drive again.

## 6,  Back to Work

At the end of January 2009 I could finally start back at work. I knew I had some things still to sort out with the hospital but I was well enough to go back part time. I seemed to be settled on my tablets, I was on 1000mg Tegretol. It was great seeing everyone in the office. They were so welcoming.

I had a timetable for going back to work that HR had sorted out. I would not be back at work full time for a while. It was nice to be able to ease myself back slowly as it was a shock to the system especially when I was already tired from treatment. It wasn't the work that tired me out, it was the getting to London and back. Still, I was very glad to start getting back to my life. In the evening I spoke to Camilla on the phone. I let her do all the talking because I was muddling up my words, I was so exhausted. I had an early night and was looking forward to a lie-in.

On Thursday the 16th January I woke up feeling like I had done a marathon. I was completely zonked out and was woken up abruptly by my mobile phone ringing loudly. It was Camilla calling me. The ring really shocked me. My parents were out of the house. My speech was really bad again. I should have ignored the phone and gone back to sleep but I never expected what happened next.

I was trying not to butt in on the conversation and tried to ignore the sick feeling in my stomach and this weird feeling of exhaustion that had come over me. I was listening to Camilla but I felt faint. I knew I should end the call but I thought it was better to have somebody on the phone because I could feel that something was going to happen. I was trying to speak but I was just mumbling. I knew I was going to have a seizure. I told her not to worry but to call my mum or 999 because something was happening. This was the first seizure I had really been aware of and I didn't know the protocol! I must have scared her – she was saying something on the phone but I was gasping for breath and couldn't hear what she was saying. I tried to lie in the recovery position, hoping that was the right thing to do. I didn't know much about seizures at that point. I was worried about things I had seen on the TV about people chewing or swallowing their tongue. Camilla was really good about it and I could hear her talking to me and saying she would get help and that she was going to stay on the phone with me. She knew I was on my own in the house so she totally took control. I could hear and feel my neck arching back. I was so scared.

I woke up and wasn't sure why I was lying on the floor. I very quickly realised what had happened. I could hear people shouting through the letterbox downstairs. I thought they were trying to break into my house, I was so disorientated. I crawled to my parents' on-suite bathroom and shut the door. I lay on the floor feeling horrible and sick. For some reason I couldn't lock the door. My whole body was in shock and my arms and hands wouldn't do anything even though I had just crawled there. I do believe that when a person is frightened they find strength from somewhere.

Now I could move more, I just took a deep breath and thought, "I'll be fine, God please protect me." I decided to go downstairs and if I needed to I would hurl myself out of the front door with all my might.

To my surprise it was an ambulance team at the door. I was so glad to see them but scared at the same time. I couldn't speak very well and I was in a real daze. They said they were going to take me to hospital to be checked out. They asked for the best number and I gave my mum's. I was so weak. One of the ambulance ladies called my mum from my mobile. I could imagine what a shock that would be for her. She was going to get to the hospital as soon as she could.

On the way to the hospital three things were going through my mind. First, was I going to lose my licence for good if this happened a lot (how silly is that? As if that was the most important thing). The second thing was the whole sitting sideways in the ambulance was not a good feeling and I hoped I would not be sick. The third thing was the question – am I going to have seizures forever? I had my priorities right didn't I? LOL. I arrived at the hospital and they checked my blood pressure and I had to have a cannula and cardiovascular examination. My mum turned up soon after me with my stuff, in case I had to stay overnight. I was so out of it. But again I was very calm and just knew God was still giving me strength, I was allowed home very quickly after my blood tests came back fine. The feeling was that maybe my tablets were not strong enough. "Oh great," I was thinking. That was all I needed. I remember asking God if I could "have a break please". I got home and fell asleep next to Mum.

I went back to the doctor on the Monday and she signed me off work AGAIN!! I couldn't believe I only made it back to work for one day. They said I needed to have more blood tests and get the seizure tablets under control. I was not myself for another few weeks. Every time I stood up I was very disorientated. I started to look into all the details about having seizures. I hated the word and I felt sick thinking about it. I thought I was going to have to sell my car for sure and lose my licence. It was horrible. I asked God many times to bless me and give my sight back – I would be patient, but please hurry up LOL!

So, I was feeling content with what I had learnt about seizures and I would be given another explanation from the surgeon who I was going to meet on Monday 26th January 2009. I just prayed for God to comfort me and give me the strength to be positive. I was surprisingly encouraged with the meeting at the hospital. The surgeon said that if I didn't have any more seizures for a year I could drive again. Sounds bad now but I was really happy with that. I was going to have to wait just a year. I knew then I may have a few seizures over the next few months.

I finally had the call to pick up my tablets and to discuss my next plan of action. My doctor told me I needed more blood tests that day. I was surprised the needles worked because my skin must have been like an alligator's skin from having so many needles poked in me. I was told I was going to be off work for another month as I adjusted to the new tablets. I was feeling really low and was not myself for a while. Everything was annoying me. I just kept to myself because I was so tired all the time. In the month I was off work I was just sleeping a lot and only went out maybe three times. I just wanted to get better and go back to my life.

On Friday 20th February I woke up early to have another blood test at the doctor's and to check my tablets were working well enough to allow me to go back to work again. I was so relieved that I was looking and feeling a lot better. I was very nervous about going back to work as the last month had been a real set back and had knocked my confidence a lot. I just tried to relax and keep positive still.

I received the results on Wednesday 25th January. My results showed my current tablets were still not doing enough for me and I had to go back up to the same tablets I was on at November (1000mg Tegretol).

I eventually went back to work again. I was only doing part time but it was hard not to worry about the seizures. I was hoping that I had had the last one now. February was going well, other than the tiredness I had. February came and went and I was happy being back to work still but I had no social life except text messages from friends. I was too tired to meet up with anyone after work or at the weekends. I started using my emails a lot and Facebook – the best tool on the internet at the time. I was able to practice typing quickly again and just work on holding a conversation.

## 7 Confidence Booster

I had gone onto a social site through Facebook on the internet. I felt like a real geek as I had never liked socialising on the internet except with good friends through Hotmail. I just assumed I would meet weirdoes. True to form, a few weirdoes found me! These were men that were not looking for just a chat. Fortunately for me at the beginning of February I met Mark. I remember thinking "not bad!" He was a Corporal in the army. I had never met somebody in the army before.

We got talking at the beginning of February. It was good for me to work at my social skills but I never thought anything would come from our conversations. Bit by bit we got to know each other and we were just being friends. He had mentioned that his mother had epilepsy and that really took a huge weight off my shoulders as he could understand what I was going through with the seizures, to some extent anyway. It was almost too good to be true. I was so nervous about telling people about my brain tumour and the seizures, I was embarrassed to talk about it as I felt like I was a weirdo and people would do that nodding thing. You know the one where they tilt their head to the side and say "oh yes I see" – when they clearly don't!

By the end of March we realised my very close friend Naomi. lived 10 minutes from Mark. Being that things were going very well, we decided to meet up. I had never ever met somebody through the internet and I had my back up in place – Naomi and her husband Andy and the dogs, in case anything happened. I am such a drama queen but better be safe than sorry I say.

I met him on April 18th 2009. I was so nervous about my seizures because knowing my luck I would have one while I was with him. The weekend went really well and we got on great. There were no seizures and I had a new boyfriend woo hoo!

The next week was going really well at work and I was coping with the travel. I had my day off on Thursday 23rd April and I planned for Mark to come down and stay with me. I was tidying up the house on the Wednesday afternoon after work. I had been fine at work and had no problems on the commute home. Suddenly while changing the bed I felt really weak. I was trying to tuck in the sheet corners and I realised I was going to have another seizure. I was so frustrated because there was nothing I could do about it. I called my mum who was downstairs at the top of my voice: "It is happening again, come here!" I just said "I'll be fine." The next thing I knew there were two ambulance people in my room again. I was taken to hospital. I didn't want to call Mark because I felt embarrassed and it was really frustrating for me and scary for Mum; she hadn't seen me have a seizure before. Mark had left messages asking if I was OK and I felt bad because I just wasn't comfortable talking about it. This was the reason I didn't want to bother with any relationship – it was just so draining. That and constantly feeling embarrassed. I really liked Mark so I decided to just tell him what had happened and give him the choice to walk away. But he was such a nice person, it didn't faze him at all.

I felt a lot better after the following weekend and I went back to work on Wednesday 29th April. Things seemed to be OK other than the tiredness. I started to open up more to Mark about the seizures and he was still not running for the hills. I was always quiet if I felt stressed or tired because I never wanted to take it out on Mark or other people close to me. Tiredness can really change a person's mood.

Mark and I went to see my friends Marina and Matthew. We got the train down. We had a really good time and Marina cooked us a great meal. We were all fine until I realised that I was feeling very unwell and felt as though I was going to have another seizure. We had to leave about 22:00 and Matthew gave me a lift home. I was getting fed up with feeling embarrassed with people – I felt like a child. They all said it was no problem but I couldn't help feeling like a burden.

After the weekend I was back on form and back to work (part-time obviously). Life seemed to be going well but I was always nervous out and about. I was falling in love with Mark. I was slowly building my confidence back up and a lot of that was due to Mark. We would always see my friends or his work friends. I think it helped that Mark was always with me. He was giving me that extra confidence to push myself even if he didn't know it at the time. I felt a bit selfish because I never wanted to meet his friends or family in Manchester but that was because I was nervous. I seemed to be fine with random people but when I needed to make an effort I would freeze and talk gibberish. As much as I liked Mark I had a few bridges to cross yet.

On Thursday 7th May the doctor had good news. My doctors and the surgeon had made a mix-up with my tablets and the combination of medication was making me really weak and very moody. Now I could get them sorted!! Or so I thought. It is crazy when you put your life into the doctor's hands only to find they make mistakes too!

I was told to relax all week and that the new tablet may set off a seizure. On the 8th and the 9th of May I wanted to see Mark and I thought I'd be fine if I just relaxed and stayed strong until I got to his place at the barracks. It took me just over an hour on the train but it felt like three hours. The tablets side-effects were horrible. It was like I was being electrified which only lasted a second but I really didn't want a full-on seizure to come on. I was so tired; I shouldn't have gone but I felt that Mark was always having to come to London and I wanted to return the favour.

I finally got to the station and got to Mark's barracks. We just watched films and TV all night with ice cream and sweets. Very healthy, I know! He had told me back in February how his mother had epilepsy so I was not so embarrassed to let him know that I could have a seizure while I was there. I was fine all night and the next day we just chilled out. I was still getting a funny shaky feeling in my body. It made me laugh because even though it is not a very nice feeling it wouldn't kill me – I knew I would be fine eventually. Mark was finishing the Cookie dough ice cream from the night before. He said I should try and sleep. I was trying but then things got weird. I was shaking all over and I couldn't open my eyes. I could hear him but I really couldn't open my eyes. I remember lying there and thinking "God, if there is any way you could intervene right now I would really appreciate it." I didn't want to have a seizure in front of Mark. My parents drove down and picked me up because I wasn't in a state to get on the train home. I really felt like a child but what can you do? LOL.

I spent the weekend resting and building my strength back up for work. I was doing OK back at work and travelling was fine. I went down to Mark's barracks on Friday 24th July after work and we had a chilled night in front of the TV. We both had people to see that Saturday night in town. He went for a work dinner and I met up with my friends Clare and Amanda. They only lived just around the corner from Mark. It was really nice to get out for the night as it was the first time I had actually had a night out even though I couldn't drink yet because of my medication. We had a really good time and the girls were always checking if I was OK. It was loud and I was shaking at the beginning but I got better through the night.

I loved being out. We bumped into Mark and his friends and they were really drunk. I didn't mind but I was a little embarrassed as this was the first time he had met my friends properly. I had a great time with the girls and I was not making a fuss over Mark. He was very drunk and started talking to every girl in the bar. The girls were asking if I was OK with it as he was being very disrespectful but, for some reason, I didn't care. I just switched off and knew that even though he was a great friend he was not the one for me. The girls went in a taxi home and I carried Mark home to his place. I got him to bed and he was out for the count. I was lying in bed and was wondering how I would tell him we should just be friends? I think it was just too soon for me to meet somebody.

The next day Mark managed to get up and said he didn't remember much from the night before which made me laugh, but something told me that that's the way he was and that he clearly liked the attention he was getting from girls. We went to the army school he worked in first thing. It was not really a school, it was more an instructor's school for the army. It was really interesting finding out what he did. I met his colleagues and friends. They were very easy to talk to. I was starting to feel really tired and I knew that was the combined effect of my tablets and being out the night before. I just couldn't cope with the tiredness. As much as I wanted a night out I knew I would be paying for it for a few days afterwards. I called Camilla to distract attention away from the fact that I was not talking to Mark or the others. I told her I felt I would have to break up with Mark but I couldn't as we were booked to go to Australia in August. I stuffed a packet of nuts in my mouth and drank a lot of water. They had a barbecue and I was straight up, first in the queue. It was the best burger ever, or was that because I was so hungry?

I went back to work the next Monday and just concentrated on my job, trying not to make a huge deal of the weekend. Mark phoned me in the week and said his mum would need to have a skin graft. I felt sick for him. It is weird because I never met her but I wouldn't put that on anyone. It is very scary and stressful for a person. It was definitely not the time to say we should just be friends. I know cancer and sickness is always with us in the world but when you have suffered from something it gives you a deep empathy with people's suffering.

I spent the next week working hard again and the tablets seemed finally to be settling down. Three weeks later we were ready to go to Sydney. I was so excited about seeing Ian and I was going to meet his new girlfriend Phoebe. The flight to Sydney completely wiped me out and I had really bad jetlag. It was so great to see Ian but no matter how much I wanted to enjoy being with him, I was too tired. My body was aching all the time. We kept really busy and I started to think that going away had possibly been a bad idea. I wasn't ready. I think that was my fault because I wanted Mark to see as much as he could in case he never came back to Sydney. I had been many times before.

Ian and Mark got on really well which was great and I got on really well with Phoebe. Ian was a great host. He got us doing LOADS, thanks Ian LOL. We went on a road trip to lots of nice places. My tiredness was starting to show. I was getting really crabby and the smallest thing was annoying me. I kept it inside all holiday as I didn't want to argue or cause a scene that would ruin the atmosphere, but I think that was a bad idea. Right near the end of the trip I just couldn't cope with the tiredness and the jetlag any more. I was boiling over with frustration. I used my hairdryer and it blew up! I laugh about it now but at the time my heart was beating so fast and the incident really shocked me. I was scared because I was worried that something like that would trigger a seizure. I just wanted to scream from exhaustion, it was as though my hairdryer had felt the force of my frustration and simply blew up!

We got back to England and sitting beside him on the plane, even though we had had a lovely holiday, I knew Mark and I were over.

One morning, back at work, my college friend Louise found me on the bathroom floor feeling very unwell. I had fainted trying to splash my face with some water. I was so glad it was her that found me. The friendships I made at Petro-Canada were truly a blessing. I was sent home and told that I just needed to take it easy and stop trying to be Superwoman! I wanted help with my tiredness and bought a treadmill to get fitter and to give me more strength in myself. I was using a Pilates video. I found it great to build me up without exhausting me. I was determined to get strong. I genuinely enjoy my fitness now. At least I know if I go out and about with my brother again I will keep up, as I will have more strength.

By November 2009 I was feeling a lot better. I went to Dubai to see Emma and George for a week. I had a fantastic holiday and it was obvious that at last I was getting stronger physically and mentally. We did so many things and I coped.

When I got back to work I was made redundant. My company had merged with another oil and gas company and they were closing down the London office in December 2009. I had known for months so it wasn't a shock. I was looking forward to using some of the redundancy money and starting a new chapter of my life.

## 8 Time to Try Again

New year, new start – again lol. In December my friend Reka said she wanted to go to Thailand and being a spontaneous person (especially when travel is on offer), I said I'd go with her if she liked. I was able to use some of my redundancy money and she was more than happy for us to go together. The next day we booked the holiday. We had two months to prepare so I needed to work on my fitness. I don't mean fit for the sunbathing and looking trimmed but I needed to work on my stamina so I could cope with a long and active holiday. It was nice to spend March away as we both had birthdays coming up and we could celebrate them there.

I was nervous, wondering if I would be feeling better by then but I hoped I would be fine. I didn't want to miss opportunities for going away or for getting a good job. I yearned to do something spontaneous, out of the ordinary. Thailand was a big challenge.

In January I was desperate to get off my seizure tablets because I was sure they were making me more tired than ever. I knew that Clobazam and Tegretol needed to be at the right level – not too high or too low. My fatigue started to get ridiculous again and I was falling asleep every day and at every opportunity I had to lay my head down. I wanted to see my seizure doctor and see if he could help. I had only a month or two until I was due to jet off to Thailand.

Then I started having really strange feelings that I could have a seizure any time. I was 26 and falling apart again. I very quickly got an appointment with my seizure doctor, Doctor von Oertzen, who said they would lower my dosage from 1000 to 800mg as the dosage I was on was higher than I needed to be on. If that worked then great but if not he would have to think of another plan of action. I would have to wait for four/five weeks to benefit from the change in dosage.

I was getting worse and worse every day. My body was just shaking all the time and making me feel so sick. I was really scared I would have a seizure and every jolt in my body made me think it was the start of one.

I knew I was not ready to work yet as things were so unstable but I was called by an agency that I had dealt with in the past and they said they had a great job doing the same type of work as I had done before with the same pay. I thought it was unlikely I would get it as

I was looking rough and could quite possibly be sick at any time. Through a determination not to give up on myself I went to the interview. When I woke up on the day of the interview I felt awful and I was not sure if I should go, but my dad said if I wanted he would come with me and wait in the coffee shop opposite the offices. It was right by Regent Street and a lovely area for shopping fans. I went into the company and sat waiting. The feeling of electricity coursing through me would not go away. I had taken up my dad's offer to come with me and so I could go straight home after the interview. I was so grateful for my parents' support. The interview went surprisingly well. I went home and went straight to bed, glad it was over.

The next day I had three missed calls from the agency. When I called them back they offered me the job! Praise God, that was just the ego boost I needed. I don't know how I got it but I guess my résumé was perfect for the job they were offering. The money was very tempting too but every jolt that went through my body was just another reminder to say, "don't take it, you won't be able to do it". I was so very flattered and I really wanted to take it. But I had to make the right decision, for me and for them, and I declined their offer later that day. They tried to offer me more money but on this occasion I knew I just would not be able to give them what they deserved and reluctantly I had to say goodbye to the opportunity.

I knew that I had to put my health first now. I had to get a more practical and achievable role while I got all my dosages sorted.

I went straight down to the doctors the next day. They said my blood test results showed that my dose of Tegretol was wrong and that led to my being sick. It took me two months to get the wrong tablets out of my system. I was really angry with my doctor for making such a big mistake. The tablets come in two types and they are identical in appearance but if you read the boxes carefully they have one small difference that makes a big difference to the patient, which my GP should know. After spending two months feeling really unwell I was back on track with the right medication.

On Monday 22nd February I went to see Dr von Oertzen. I had been very lucky to be able to use the private insurance through Petro-Canada and even though I had left the company I was able to use a number of benefits until March 3rd 2010. Doctor von Oertzen was happy with my progress and said that I could come off Clobazam. However, as I had experienced before, everything has withdrawal symptoms – although these where not as bad as before as the dosage was reduced very slowly. The worst part was the sleepless nights and being so tired. It was making me feel miserable and sorry for myself. The side-effects for most seizure tablets include drowsiness, dizziness, decreased appetite, shaking, confusion, unsteadiness, headaches, loss of memory and double vision. Sounds lovely doesn't it?

The doctor said that as I would be on only one tablet after the 3rd March I could drive again and I could have children if I wanted to as well, although not necessarily at the same time! I didn't want the children, but it was nice to know although I have found that it always helps to have a man first! It was great seeing the bars come down from my jail. It really had felt like a jail sometimes.

On Monday 1st March I woke up feeling very, very sick. This time it was not from feeling ill, just with butterflies as I was going to see Dr Minhas my surgeon. It had been six months since I saw him last. All the way to the hospital I just wanted to throw up. It might seem silly but when you have all but lived in a hospital with all the appointments and the operations you can't help but develop a natural feeling of anticipation, fear and worry.

I started to relax a little when, thank God, Dr Minhas said he was happy with my MRI scan. However, even though he was happy he said there was a very small light part on the X-ray that might be more than it should be. He said he was not hugely worried, but he did say that although normally the next scan would be in a year, he would like me to have an X-ray in three months' time. That was a bit of a blow, and I knew I could not rest until I got a totally clear result. I had had good news and most people would think I should be happy with that, but for me only an all clear would put my mind totally at rest.

It was March 2010 and I was settled on the tablets and the scan was positive and I could relax slightly. Finally I was ready for my trip to Thailand with Reka. We went to Bangkok, Phuket and Ko Samui. We knew it would be a great opportunity and I knew I needed to get out and about to feel more confident. I know most people would probably not take such a big step so early, but I have never done things by half. It was a very relaxing holiday, not too strenuous. We were up early and went to bed early. We were real partygoers – not! I saw the most amazing things. I was not always very forward with buying things or asking for help as my speech was not great and I was worse when I was panicking. I let Reka do that part as I didn't want to risk muddling my words up and any maths skills I might have had had gone right out of the window since the surgery. We returned home on the 18th and it took me a whole week to get over it. The exhaustion was worth it though.

Once I had settled back in at home a lady from Petro-Canada's insurance company came to meet me and said she would still help me get back to work again with a company that would understand and be sympathetic to my situation. It was really nice that Petro-Canada wanted to help me even though I didn't work there anyone. After talking through my history she said she would be in touch in about a month with some ideas for my future. I explained my recent visit to the hospital to her and the fact that the consultant thought there was a small chance that the tumour was still there but I so wanted to go along with the idea I said I was sure it would be fine.

A month went by, and the lady from Petro-Canada came to see me a few times in May to sort out my going back to work. She would look through the options for jobs. If I am being honest, she was no help at all, sorry, but it is true. I could see that I was going to have to take charge of my life and not rely on others to give me the help I needed.

On Tuesday 4th of May I had a phone call from the recruitment lady I had been dealing with in January. Although I had not been in a position to take the role in January I was surprised and happy when she came up with a new role as a PA in a London Bridge company. I wasn't sure if I was going to be strong enough to take it yet. A busy PA role might be a little full-on seeing as how I was not a quick as I had been. I needed to think about it hard as I didn't want to make a wrong decision. Going for a job and then not being up to it would undoubtedly do me more harm than good. The lady I was dealing with, from Petro-Canada said she thought it was a bad idea and I shouldn't go for the job. It was hard walking away from the opportunity as I had always worked, and not working felt like failure. It made me feel like I wasn't good enough but, maybe this was something I was going to have to get used to.

The next month was all a bit of a blur.. I was looking for work but was conscious that I wasn't going to settle on somewhere I would not be happy in. In July I had an eye test to see if they had improved and my peripheral vision was better. I was so nervous when I had it. I already knew when they gave me the results they weren't good enough to drive again. I was hitting thing like tables and people all the time because I couldn't see them. I received a letter from the eye clinic who confirmed that I'd never drive again. It was really hard. I felt so sick for days – I was shaking. One of my absolute joys was driving. I cried for days and even to this day I feel sick when I think that I will never get to drive again. I don't think I will ever be happy with not driving. I think it was the waiting that was hard. I still think about it when I really feel like just getting up and going on a road trip but I have gotten used to using the train now and it is not the end of the world. I just have to ask a friend to drive if I want to go to somewhere I can't get the train to.

Finally on the 9th July 2010 I got a much more positive letter – the MRI scan showed that I was all clear. I did not need to be scared anymore. I would not need another X-ray for a year, such a relief.

At the end of July, Emma and George were getting married in Poland. I was nervous, as it was my first big event since before the tumour was removed. I was going to be a bridesmaid! My worst nightmare was being in front of people I didn't know (well some I did, obviously). I had about five panic attacks over the whole thing but I was still determined to do it and felt very lucky to have been asked to be a bridesmaid and celebrate the special day. The wedding was in Poland as George is Polish, and they were getting married in beautiful Krakow. The whole trip and the wedding went really well and I had a few Coronas/glasses of wine to help with the nerves. I was glad once it was over but only because it was totally out of my comfort zone. The wedding was beautiful.

When I came home I was trying to plan how to get back to work and have some sort of career plan again. Unfortunately because it had taken so long to really get back on track again with work my confidence had feeling suffered? Finally in September I was asked by the job centre if I wanted to join another rehabilitation centre called Attend ABI (Acquired Brain Injury). I immediately jumped at the opportunity, as I would take any help to get back to where I had been before all this. Attend ABI is a service that helps people with an acquired brain injury back to work. They provide training and support to help clients back into volunteering, education and the workplace. It shocked me, as they want nothing in return but want only to help. They are part of the 'Prince's Trust'. Attend is designed for adults of working age who have an acquired non-progressive brain injury. If you have substantially completed your medical rehabilitation or need additional support to return to work, then Attend ABI can help you. The programme provides you with specialist neuro-rehabilitation services to help manage the cognitive, behavioural and emotional effects of brain injury. The programme includes specialist occupational assessment and planning services, and structured 'real world' work placements, as well as rehabilitative support both inside and outside of the workplace.

On the first day I joined Attend ABI, the 1st November 2010, I was feeling very positive but very nervous. They were based in London near Oxford Circus, in a beautiful building very near to Harley Street. I immediately liked the people I was learning with, and the staff. I was the only girl in a class of eight. I'd be lying if I said I didn't like all the attention. We were all around the same age but from totally different backgrounds.

The course lasted for three months. During December I went to spend Christmas with Emma and George in Dubai for a week and then flew over to Sydney to see Ian for New Year's Eve, courtesy of the tax rebate I had received. I wanted to push myself to get out and about on my own and build up my confidence which seemed to be and endless battle.

I had a lovely time. It was very relaxing and I really appreciated the time away. Once I was back from my holidays I went back to Attend ABI for a few more weeks. It was time to put everything I had learnt together and then start looking for work again.

I really had learnt a lot from Attend ABI. The things I learnt from there and from the other people on the course were so helpful and helped me a lot.

I learnt that it was OK to need a lot more rest than I was used to. I learnt not to beat myself up for being 'lazy'. Brain trauma from any cause leads to physical fatigue. It takes a lot of energy and it is very tiring for our brains to think, process, and organise.

I learnt that energy levels fluctuate, and that even though we may look good or seem to be 'all better' on the outside, that might not be the case on the inside! Cognition is a fragile function for a brain tumour survivor. Some days are better than others. Pushing too hard usually leads to setbacks, sometimes even to illness.

Brain tumour recovery takes a very long time; it can take 10, 20 or more years. It continues long after formal rehabilitation has ended. And it can be exhausting trying to live up to the heartfelt desire of those around us who are expecting us to be back to exactly who we were, because, superficially, we look better.

We are not being difficult if we resist social situations. Crowds, confusion, and loud sounds quickly overload our brains, which don't filter sounds as well as they used to. Limiting our exposure is a coping strategy, not a behavioural problem!

If there is more than one person talking, we may seem to drift. That is because we have trouble following all the different threads of conversation. It is exhausting to keep trying to piece it all together. If you live with a brain injury patient or a brain tumour survivor, try to notice the circumstances if a behaviour problem arises. Our 'behaviour problems' are often an indication of an inability to cope with a specific situation. We may be frustrated, in pain, overtired or there may be too much confusion or noise for our brain to filter.

Please have patience with our memory. Keep in mind that not remembering does not mean that we don't care.

If we need to do tasks the same way all the time, it is because we are retraining our brain. It's like learning main roads before you can learn the shortcuts. Repeating tasks in the same sequence is a rehabilitation strategy.

If we seem sensitive, it could be a result of our brain tumour or its removal, and it may be a reflection of the extraordinary effort it takes to do things now. Tasks that used to feel automatic and take minimal effort now take much longer, require the implementation of numerous strategies and are huge accomplishments for us. Be patient with us! Don't confuse hope with denial. We are learning more and more about our amazing brains and there are remarkable stories about healing in the news every day. No-one can know for certain what our potential is. We need HOPE to be able to employ the many, many coping mechanisms, accommodations and strategies we need to navigate our new lives. Every single thing in our lives is extraordinarily difficult for us now. It would be easy to give up – without hope.

The Attend ABI course finished in February 2011. Through their support I started to volunteer at St Christopher's Hospice in Sydenham. It was about 30 minutes from my house. I was there every day for around two months. It was strange at first being in a hospice. It was sad to think that the people there were often at the end of their lives. I was helping on the 'fundraising side'. I couldn't answer the phone, and they understood my fear of the telephone after a seizure back in 2009 was caused by a phone call. My role at the hospice was looking for businesses that could donate prizes for the various charity events they held. I loved working at the hospice. It gave me a purpose and it felt good helping others.

Although everything was going well at the hospice, I knew I still needed to find a paid job. I got in touch with a recruitment agency and got a few interviews in London. It was not going well as I was getting interviews but for one reason or another the jobs offered were not appropriate for me anymore. My CV screamed 'Admin/PA' but that was not realistic, I just would not be able to do justice to that typically very busy role, I knew that. I went to some agencies to see if there was something a bit easier as I needed to ease my way back into work full time and not overdo it after this big gap in my CV.

## 9.  Time to Ask for Help Again

The volunteering at St Christopher's Hospice was great I was working 7hr days but working every other day so I could always rest in between. As it was in the fundraising office it was not too demanding. After a few days I found my sleep was beginning to be very restless again. At first I thought it was because I was spending too much time thinking about my life and wanting to work – wondering constantly where I would go from here, where I would get a paid job and what kind of role I could realistically take on. I was getting further and further away from what I had been like before my operation and I was looking for answers again. I was starting to find it hard to pray or even have an understanding of what I was going to do about my future. It is not a nice feeling when you have no goals, or at least when you can't even recognise a goal for yourself that you could realistically achieve.

I was having severe panic attacks in the day and even more at night. I think it was from my constant worry about life and what I would do. The attacks would always feel like the start of a seizure. They weren't, but it is hard not to have the fear of seizures in the back of your mind after you have had experience of them. And although I am not epileptic with flashing/strobe lights I always felt as though I should not look at them as I felt they could easily trigger a seizure. It was suggested I should see a psychic to see if he or she could do a reading to make me feel at ease and to reassure me that everything would be OK with my future, but that was not a good idea for me. I have never given much thought to that kind of thing and never really wanted to get involved or entertain the subject, as it is not something I believe in. It is strange how you will go to a great extent to find answers.

I was given a number for a popular psychic and even though every part of my body said "Don't do it", I did. Maybe it was me rebelling against my own faith as things were going wrong again and I was getting depressed with all the ups and downs of my recovery.

So I spoke to the psychic. She said all good things and told me that everything would be fine and gave me some more information. What she said was enlightening but after I had the reading my nightmares and paranoia got worse and worse. I was constantly feeling there was something in the room with me. The lady I spoke to said it was my spiritual guides and they were there to help me with my decisions. That really freaked me out.

For weeks after I was afraid of the strange feeling I had like something was always there. I wished the lady didn't say I had a spiritual guide with me always. I prayed for forgiveness and I asked my Christian friends to pray with me over the situation. Soon the bad feeling had gone but I was still struggling with where I should go from here.

Roland had told me to look up a particular sermon by pastor Jesse Duplantis, who is one of my favourite pastors from America. The sermon was about how much God loves us and what he has waiting for us one day and how he wouldn't let anything happen to me as I am his child. I felt better already.

I sent Jesse Duplantis an email to pray for me and explained my situation. A lady called me from his ministry and then she prayed for me on the phone. Two days later I was given the opportunity to interview for my friend Amanda's company. I thought "Wow that was quick, thank you God." I got a second interview and I got the job! I started the next week on the 28th of March, the day after my 28th birthday.

The job went well for the first few weeks and I was getting on with everyone at work. I loved being back in the city. Getting up early and just being back to how I wanted to be, struggling into work along with all the other commuters, full of purpose for the day ahead. I knew the job was not easy but the fact I had Amanda there gave we a boost to try and I was able to speak to her if I got stuck with anything. Can you believe I even met a guy at the company and we went on a few dates. The relationship wasn't going anywhere but at least I could see I was still in the game and had my swag! LOL.

Into the fourth week of my new job I realised I was working nine-hour days and that is not including the travel time there and back. So I was actually doing eleven hours a day. I liked being busy but my body was struggling; I was coming home and going straight to bed and then getting up for work. Now Amanda was getting married in April and I was really excited, as I was to be one of her bridesmaids. I knew this was going to be another trial, as I would have a very long and busy weekend and then have to get straight back to work.

The wedding was amazing and we had a lovely time. Amanda looked stunning. The night before the wedding Amanda, to observe tradition and not see her groom, had slept in my room. She said she noticed I was shaking in the night – my feet and legs were shaking. I had heard that people do have seizures at night and that this kind of shaking was a sign.

Monday came around all too soon, and it was time for work. I was so exhausted I couldn't even remember my name or what to say on the phone. I was forgetting messages. I was not with it at all. I felt like I was in a bubble that I needed to pop to get back out into the world around me.

It was April and on the Thursday of that week we were all looking forward to the long weekend – the big royal wedding for Prince William and Kate. All the UK had a four-day holiday. I needed it badly to get back some energy. When I went back to work I spoke to my supervisor. We talked about the situation and we agreed it was not working. I should have been devastated but I was glad. I had tried, and the experience had made me realise that no matter how much I wanted my life back to how it was, what I was trying to do was not realistic at all.

I went back to the hospice on the Tuesday. I was tired but I was doing OK, and then a week later I got flu again – go figure. I am a walking advert for lemon and honey! I knew I had really overdone it with the job in London and should have had a rest before I went to the hospice, but I just wanted to keep busy so I did not get too comfortable being idle...

I really loved working at the hospice; we were setting up for another fun run. The day after the fun run we had to pick up all the water bottles that were left over. They were being looked after in Waitrose supermarket as it was close to where the fun run had been held.

There were only three of us and there were hundreds of boxes with eight bottles in each. I have never claimed to be a manual person. I can just about lift my head these days let alone lift boxes. After having got all the boxes on the van we went to drive off and bang, the tyre blew. I laughed lol, it shows how heavy the cargo was. We waited around for three-quarters of an hour until the AA turned up to change the tyre. Back at the hospice we had to take all the boxes back off the van and put them onto some trolleys. I was fine doing stuff but knew this wasn't the great thing for me to do with so little energy.

I finally got home and went straight to bed. It was one of those moments when you hit the pillow you are out like a light. I woke up about 09:00 and was feeling very tired – my body was aching. I went to go and wash my face and locked the door. I knew Mum was just outside the door. Next thing I knew I was back on the floor wondering "Did I have a seizure or did I faint?" I was OK, and was certain I had just fainted, nothing more. I just went back to bed and slept it off. I relaxed the whole day and then had an early night. I woke up early the next day at 05:30 as Mum and Dad were picking Ian and Phoebe up from the airport. They had had a long flight from Sydney and I knew we needed to keep them up for the day. I did get up at 05:00 to say bye to my parents and was trying to stay awake but I was still really tired from the bottle carrying. I said hello to Ian and Phoebe and told them I would have to just get some more sleep and then I would be more awake later.

I finally woke up and had a catch-up with them. It was so nice to see them. In the afternoon we went shopping in town and bought a game for my Wii. It was another guitar (Wii guitar). We went home and played on the games. It was good but later that day I started really feeling my shoulder aching terribly. I realised it was the side I had fallen on. It was depressing at times when I wanted nothing more than to keep up with Ian and Phoebe while they were with us, and instead I had to keep resting. While Ian was home we celebrated his 30th birthday. We had a huge family party for him. I had to keep taking rests in the day and went for a nap every now and then as I found it very loud with everyone in the house.

After the party I was pretty much bed bound relaxing, my shoulder was getting worse. While I was home I had to do something to stop myself going mad from boredom and worrying about myself. I hadn't really talked about my actual experience with other people very much. I had always kept this to myself and I had never looked into what a brain tumour really was. It had been three year but it was such a scary subject to me that I adopted a sort of head in the sand attitude.

But I felt a lot stronger in my mind now and I finally wanted an understanding of brain tumours. I started looking at websites. I knew my situation but had never actually spoken to other brain tumour survivors or sufferers. I had met people at the Attend ABI but they were all people who had had head injuries rather than brain tumours.

After looking at the various websites that were out there, I realised there were also forums on Facebook. I went to an American one first for meningioma patients. As I started reading the comments people were writing I began to cry. It was very hard reading about people's experiences and their struggles. It was nice to see the problems I was having were very normal. 'Normal', what on earth does that word mean?

I could see how hard people were finding their situations and I wanted to help. I think it also took my attention away from myself and tempered much of my introspection, which was a good thing. Somehow I did not want my family and friends to hear me talking about my brain tumour journey and recovery because, as ridiculous as it sounds, I think I was embarrassed. It was as though it was my fault or that I was an embarrassment, with a disability. That seems so silly now but I didn't meet anyone else in my position for three years after my surgery. After a few days of looking through the websites and forums on Facebook I decided I wished there was a more positive website that really focused on cheering people up.

It was May 20th in 2011 and the anniversary of the day I had been to the optician three years earlier to find out that my eyes were bleeding at the back. It felt like a good day to start a new chapter in my life.

I set up a page on Facebook and I called it 'Aunty Meningioma'. I did not want to give my name away at that point as I didn't want everyone to know who I was and didn't want my friends or family to see my true feelings about the whole thing. I also wanted to speak with others and didn't want my age to put people off. It was easier to make up a name and I thought people might find the name 'Aunty' comforting.

Every morning, afternoon and evening I would put an encouraging but funny photo on the page to make people smile. Day by day more people were adding themselves to the page. People seemed to be really benefiting from my positive comments and they were always related to brain tumours. It felt good.

Ian and Phoebe were still over and every time somebody added to my Facebook page I would say out loud '1 more yay'. I think they all thought I was being a little strange and people wanted me to move on but in my mind this was just the beginning of my recovery and finally meeting others. Better late than never I say. Ian and Phoebe returned to Sydney after their three-week holiday with us. It was sad seeing them go but I was so glad they came to see us.

Once they had gone back to Sydney I was still having trouble with my shoulder. It was very painful. One night it got so bad I almost went to casualty to get some help. It was only through the fear that I would not be able to cope with the noise and confusion in casualty on a Saturday night that I decided to just put up with the pain, and wait until Monday to go to the doctor.

I went back to see the doctor and mentioned that my neck was hurting from falling the week before and it wasn't getting better it was getting worse and I was having terrible spasms. I thought it would be a good idea to mention that Amanda had noticed me shaking in the night when I stayed over on her wedding night. The doctor gave me some cream to help with my shoulder and he said I should ask somebody to sleep in the bed with me and see if I did it again – so with no takers on being my bed buddy of the male variety, I asked my mum to sleep with me! She did it for three nights and on each of the three nights I did the same thing, shaking. The GP doctor said I should speak to my seizure doctor, Dr von Oertzen. He was away on holiday so while I was waiting on Doctor von Oertzen I decided to send him an email to voice my concerns.

*Dear Doctor von Oertzen,*

*I know I have an appointment on the 27th June 2011 but I am concerned. I have been having many problems again with lack of sleep and feeling very, very tired all day. I am used to the tiredness but now it has gotten worse. I have been struggling with my speech again, with finding words and with focusing. I find any kind of noise is hard on my nerves; this has been going on for a month. I have also been getting very tense in my shoulders and feeling very sick, dizzy and disorientated. I was not like this for a good two years but it is very unsettling. This was happening before I reduced my Tegretol Retard from 1000mg to 800mg. I thought the reduction would help but it is just stayed the same and I am feeling very sick and nervous like I may have a seizure because of the tiredness and stress. Please help?*

*I have spoken to my GP already and as usual they just did a blood test which was for anaemia and Tegretol levels which were normal, but I know my own body and something is wrong. I never make a fuss normally but I am nervous and it is not nice to have been feeling sick and weak for nearly a month.*

*I look forward to hearing from you.*

*Claire*

After my email Dr von Oertzen set up a meeting for me with him. We met at St George's with other people observing us. I was feeling very low and had to really push myself but I wanted some help, and how would they know what I needed if I couldn't speak up? Doctor von Oertzen said he wanted to wait for my MRI results to come through as I was due to have a scan any day – that way he could dismiss the possibility that the problems I was having were tumour related.

I had the MRI a few weeks later and had an anxious wait for the results. It then took three weeks to hear the results as the hospital had misplaced my file. Once I met Dr Minhas again he said he was really sorry for the wait – I should never have had to wait so long to hear the results of my MRI. But it was all good news and the MRI was fine. It was so nice to see a clear picture.

Dr Minhas said my problems may be arising from seizures. He thought we should set up an electroencephalogram (EEG). He said he would speak to Dr von Oertzen and discuss the next course of action.

A week later I got an official letter that said my scan was all clear and that there were no more problems from the surgical side. It went on to say that I might have depression. I was really hurt by this because when I had met Dr Minhas the previous week we had agreed it was not depression and that the small seizures might be making me feel sick and very tired. I felt sad all weekend because I couldn't understand why he was now saying that my problems were due to depression. Maybe he was right and I was just going mad.

Finally I had a letter from the hospital again about the discussion with Dr Minhas. They said the letter was wrong and that it shouldn't have said I had depression. It was a copy of the one they had sent me six months earlier but it was mislaid, so they sent it so I could have it in my files!

It did say they wanted to look into my seizure medication and also that they would like me to have a few tests for my cognitive abilities. I then went to have some tests and the results showed I definitely needed some cognitive therapy. They also set me up another appointment to see Dr von Oertzen but he was on holiday so I saw another seizure doctor. She was a lady I had never met before. She said from my symptoms that my problems were definitely due to my seizures. She said I should go from 800mg Tegretol to 1600mg in a gradually increasing dose. They would set me up with a psychologist and a cognitive therapist.

While I was happy we had a result I was sad that it was going to take time to stop experiencing that feeling in my body that a seizure was imminent. She told me to start a diary to track how I was feeling, to see when the seizures were occurring. I wrote the diary and it seemed every night I was struggling with tension and funny feeling of falling,

Waiting for the tablets to kick in again took time. It took 10 weeks to get on the correct dosage and I was still having very strange sensations and was waking up in the night shaking. I knew I had to hang in there, I knew it would start to get better soon.

I had got to 1400mg and it was not working, and my seizures continued day and night so it was not a great existence. I was waiting to see a shrink so I could talk about my problems as I was not leaving the house for fear I might end up jumping around in the street. I was becoming a recluse.

I didn't go out for over six months while the tablet dosage was still going up. I felt so unwell and tired it was absolutely exhausting.

On Friday 19th of October I had a meeting with Doctor von Oertzen who was not happy with the previous doctor's course of action four to five months earlier. He said I had been put on the wrong tablets and I was not having seizures, it was just vivid dreams! which would have set of anxiety issues.

I spent from October to December adjusting to new tablets called Lamictal (Lamotrigine). Slowly! I was having endless problems with feeling sick, headaches, sickness, vision and low weight (the latter not necessarily a bad thing!).

I carried on with my facebook page while I was stuck home still, I really learnt a lot from the people who shared their brain tumour experiences and although they were seeing my pictures and posts they would explain what their connection to a brain tumour was. At that point I did not know enough about brain tumours to realise there are actually 120 types of brain tumour out there and that they are graded.

I had nearly 600 people on the page by September and realised maybe I should change the name as I didn't want to make it just for 'meningioma sufferers/survivors'. I met Keith Davies, a fellow brain tumour patient who had a tumour called a parasagittal atypical meningioma. He was a similar age to me and after sharing our brain tumour experiences he was very helpful and we came up with the name Aunty M Brain Tumours. We devised a catchy introduction to tell people what I did and to make them feel very welcome.

**Aunty's Wall:**

**"Join Aunty M who is all about positive thinking and is a place to meet friends who understand about brain tumours. The thing we all seek in life is to get through the hard times and keep smiling. I was diagnosed with an intraventricular meningioma in 2008. I have been through it and still am going through it but I am on my way out the other side now and I am staying positive!**

Eventually in December, I ventured out and turned the laptop off. I was so paranoid that I might have a seizure. Although I had finished the swap-over with the tablets, I was still having trouble with anxiety and worry. Unfortunately the one time I did go out to do some Christmas shopping I caught a cold – typical.

I went to meet Dr von Oertzen again to reassess my situation and what we could do, going forward. I was feeling nervous as I had been hiding in the safe cocoon of my home. Dr von Oertzen was happy that I was going to get counselling and said we would see how I got on with that.

When I left the appointment I was not feeling very confident as I was told I would be on seizure tablets my whole life. He said my hippocampus had been damaged in the surgery and there was a lot of scarring that would trigger seizures. Being me and being in central London, Mum and I went to a few shops to browse and try to regain our equilibrium after the rather depressing news. I bought some nice things and managed to forget about the meeting for a short while. Back on the train and back to West Norwood, we got the bus up the hill. I was listening to my iPod and then it hit me. Damn! I won't ever get off the meds! I glanced over and saw a person reading a newspaper. On the back it said in big letters

'SHAME'. That was just what I wanted to see! It would have been funny if it weren't so tragic!

I got home and felt bad for a few minutes and then squared my shoulders and decided I would just have to get on with it.

Doctor von Oertzen reiterated to my GP that I needed to go and do some cognitive counselling and some counselling specifically for people who have had brain surgery. I was feeling positive about that. I didn't sleep all night. I started watching some TV and then suddenly on the news they were talking about 'cognitive behaviour'. They were explaining everything and I was just thinking "Thank you Lord, good timing."

I understood now it was not about being slow or just strange. Cognitive problems were like a person who has had their mind set put into the wrong video machine. All I needed to do was to get the right one back to carry on!

It is not a quick process. It takes a long time but with determination, I know I will get there.

I spent Christmas on my own. I was meant to go to my uncle's house for Christmas Day but my cousin had just been diagnosed with Hodgkin's Lymphoma so it was not a good idea to spread my germs to him. I was more than happy on the sofa with a lot of feel good movies and my lemon and honey drinks. Thank goodness for 'Sky Movies'!

Finally I got to talk to my GP and got her talking with my seizure doctor's assistant, as Dr von Oertzen was away on holiday. I think they felt sorry for me as it was Christmas so they made an effort to sort me out, which I really appreciated.

They said there must be some other reason for my symptoms. I spent the night feeling so unlucky, only to be reassured the next day that all my symptoms were not from my seizures. I was having panic attacks and anxiety attacks. I then had to speak to my GP and set up a meeting with a counsellor. As it was Christmas, I had to wait until after the New Year celebrations.

After working tirelessly on Aunty M Brain Tumours through Christmas I decided the blog etc. was working well and within 15 months I had over 13,000 people connecting on my different websites.

On December 31st I was going to spend New Year's Eve with me, myself and I, but I was glad I had a plan. Ding Dong! It was January 1st 2012. Can you believe it? I slept through it as I was fast asleep by 22:00 lol. It was nice to speak with brain tumour warriors around the world on the 'Aunty M's facebook page. It was not such a lonely Christmas as there were others in the same situation and we were all there together to say goodbye to the year and bring on the new one.

To Be Continued

## About the Author


## A Brain Tumour's Travel Tale

Claire Bullimore was born 1983 in London, daughter of Eileen and Patrick Bullimore and sister to Ian.


She was diagnosed at the age of 25 with an Intraventricular Meningioma, in other words a brain tumour!


After her own battle Claire set up a free supportive social network called Aunty M Brain Tumours. It is for individuals and families affected by a brain tumour.


Website:

www.auntymbraintumours

Printed in Great Britain
by Amazon.co.uk, Ltd.,
Marston Gate.